A series of student texts in

CONTEMPORARY BIOLOGY

D0989218

General Editors:
Professor Arthur J. Willis
Professor Michael A. Sleigh

Statistics and Experimental Design

Geoffrey M. Clarke
M.A., Dip. Stat. (Oxon.)

Reader in Statistics
Mathematics Division, University of Sussex

Formerly Statistician at the Long Ashton Research Station
and Lecturer in Statistics, University of Bristol

Distributed in North America by
UNIVERSITY PARK PRESS
233 East Redwood Street
Baltimore, Maryland 21202

© Geoffrey M. Clarke, 1980

First published 1969
by Edward Arnold (Publishers) Ltd.
41 Bedford Square,
London, WC1B 3DQ

Reprinted 1971, 1973
Reprinted with corrections and additional exercises, 1976
Reprinted 1977
Second edition 1980

Paper edition ISBN: 0 7131 2797 X

Printed and bound in Great Britain at
The Camelot Press Ltd, Southampton

Preface to the Second Edition

In the ten years that have elapsed since this book was first written, statistics has found its way into a number of biology courses at pre-university and college levels. When students reach university or college courses, however, there still appear to be the same needs and problems as in the past. Most students of biology do little mathematical or numerical work, and find difficulty in thinking in symbols, so that they need to learn statistical methods in an 'applied' way, seeing the use of the methods and the reasons for them. In fact this is a very good way of learning statistics, because many more gross errors in applying statistical methods arise from not knowing when they work than from getting the arithmetic wrong (though there are enough of these, too!).

There has therefore seemed to be no reason to change the original plan, of aiming to present the ideas behind, and limitations of, standard statistical procedures as well as the methods themselves. It must be admitted that people sometimes appear to find this logical thinking about the methods just as hard as understanding what the symbols in the standard formulae mean; but there is no substitute for thinking before using *any* scientific method, statistical or otherwise. In biology it is often possible, with suitable forethought, to plan to a fairly detailed extent the collection of data and the conduct of experiments, and so emphasis is placed in this book on the elementary aspects of experimental design. To study the basic principles of experimental design and analysis helps in understanding some of the commonly-used tests of significance, which were designed for such situations originally.

Some changes have been made in this edition. The question 'Which is the right test for my data?' is often asked, and when it has been sharpened up by thinking what features of the data or the problem are the important ones it can be answered: an attempt to do this is made on pp. 157–8. It is hoped that this will save readers much time in locating the methods as they need them. Additions to Chapters 10 and 17 increase the material on nonparametric methods, though by no means to the level of a catalogue of all the Thirty Nine Steps (or is it Fifty Seven Varieties?) in the nonparametric world; there are of course well-defined problems in which such methods are essential, but not so many of these problems are important biological ones. In experimental design, factorial schemes frequently have factors at more than two levels, and a new section extends the original Chapter 15 on this topic.

Worked Examples are included in the text; and there are answers and comments, collected at the end of the book, for the Exercises that follow the chapters. A student working on his own would be well advised to attempt all the Exercises before moving to a new chapter.

Many of the principles of elementary statistical methods apply quite generally over a wide range of sciences, and it is hoped that the book may be of use to scientists in disciplines other than biology, provided they do not mind the examples being from biology or biochemistry—no great depth of knowledge in these subjects is required to appreciate them!

I am indebted to the Biometrika Trustees for permission to reprint parts of Tables 8, 12, 13 and 18 from *Biometrika Tables for Statisticians*, third edition; also to the Literary Executor of the late Sir Ronald A. Fisher, F.R.S., to Dr Frank Yates, F.R.S. and to Oliver & Boyd Ltd., Edinburgh, for permission to reprint parts of Tables III, V and VI from their book *Statistical Tables for Biological, Agricultural and Medical Research*, sixth edition.

I would like to acknowledge the help and encouragement received from Professor Arthur Willis, General Editor of the series, and from the Publishers, when I first wrote this book, and also their continuing interest and the various comments which I have received from users. All these have, I hope, helped to reduce the number of errors or obscurities that might otherwise be present. Comments will continue to be welcome.

The University of Sussex   G. M. C.
Falmer, Brighton, 1980

Table of Contents

Preface   v

Symbols   ix

Introduction   xi

1 Populations, Samples and Variates   1
2 Summarizing Observed Measurements   7
3 Distributions and their Characteristic Properties   12
4 Probability and the Binomial Distribution   23
5 The Poisson Distribution   32
6 The Normal Distribution: its Occurrence   37
7 The Normal Distribution: its use in Significance Tests   44
8 Other Tests of Significance   58
9 Setting Limits to Estimates   77
10 Correlation   83
11 Linear Regression   92
12 The Principles of Experimental Design: the Completely
   Randomized Design   101
13 Randomized Blocks and some extensions   114
14 Precision of Results   124
15 Simple Factorial Experiments   128
16 Missing Observations and Non-Normality   142
17 Nonparametric Significance Tests   149

   A Guide to Choosing a Significance Test   157
   References and Bibliography   159

Answers to and Comments upon Exercises   160

Tables   177

Index   184

Symbols

r_i = a typical value of a discrete variate r
x_i = a typical value of a continuous variate x
f_i = the frequency with which x_i occurs in a sample
$\bar{x}$ or $\bar{r}$ = the mean of a sample of observations
N = the number of observations in a sample
M = the median of a sample of observations
s^2 = the variance of a sample of observations
$\mu,\ \sigma^2$ = mean and variance of a probability distribution
$\mathcal{N}(\mu, \sigma^2)$ = the normal distribution whose parameters are μ, σ^2
$n,\ p$ = the parameters in a binomial distribution
q = $1 - p$
$\hat{p}$ = estimate of p calculated from a sample
λ = the parameter (mean) of a Poisson distribution
$\hat{\lambda}$ = estimate of λ calculated from a sample
N.H. = Null Hypothesis upon which the calculations of a significance test are based
A.H. = Alternative Hypothesis, accepted when N.H. is rejected
d.f. = degrees of freedom
$t_{(f)}$ = Student's t distribution with f degrees of freedom
$\chi^2_{(f)}$ = the χ^2-distribution with f degrees of freedom
$F_{(m,\ n)}$ = the F (variance-ratio) distribution with m and n d.f.
ρ = correlation coefficient in a population
r = estimate of ρ calculated from a sample
r_s = Spearman's rank correlation coefficient
b = slope of regression line
$\hat{b}$ = estimate of b calculated from a sample
d = the unit (standard) normal deviate (often also called z)
S.S. = sums of squares of deviations about the mean
M.S. = mean square (S.S./d.f.)
$\hat{\sigma}^2$ = estimate of variance calculated in analysis of variance
e_{ij} = error term in a linear (analysis of variance) model

Introduction

Experiments in the physical sciences often aim to estimate a numerical *constant*, and a student whose estimate is far away from the known true value has made an error at some stage of the experiment. Perhaps the error arises in using the equipment, or in not allowing for some change from normal environmental conditions when carrying out the experiment and the calculations involved.

In biology, a further important factor must be recognized. This is also true of other subjects as widely different as industrial experiments in engineering, metallurgy, textile production or pharmaceutical chemistry; so statistical methods developed for one subject very often apply to another. The all-important common factor is that individual items of material, individual plants or animals or engineering specimens or units of textile, *vary among themselves naturally even when they are treated alike.* Two plants, grown originally from the same batch of seed in identical pots side by side on a greenhouse bench, and given the same amount of fertilizer and water over the same period, will not grow to exactly the same height; nor will twin animals of the same sex, living in the same cage and receiving the same diet, put on exactly the same amount of weight. ***Experimental Error*** in this sense (a universal but perhaps unfortunate phrase) means the natural variation which is present among the individuals or units concerned, even when they are treated alike and are in identical conditions.

Now suppose that we examine a considerable number of individuals. We may measure the heights of many plants, or we may weigh many animals of the same type and so collect a whole set of height or weight records, from a whole ***population*** of individuals. When looking at the natural variation present in these records of height or weight, we often see that it has a pattern: for example, some plants are much taller than the general average, some much shorter, but the majority are fairly near this average. Biological data can often be explained by one of a few standard statistical patterns of natural variation—*distributions*, as these patterns are called: thus we shall speak of the distribution of height through a population of plants, meaning the way in which height varies when we measure a number of similar plants all growing in the same environment and conditions.

In order to make sense of a whole population of records, we often try to see the pattern in them by drawing a suitable diagram. Also we shall want to summarize the information about the whole population by calculating suitable numbers: a good method of summarizing a large collection of measurements of heights of plants is simply to calculate their *average* height. Frequently, in experiments, two populations of plants will be exactly comparable except for one factor: for example, one may receive a standard fertilizer treatment while the other has this plus additional nitrogen, growing conditions and environment being otherwise very similar. We shall want to compare these two populations, and it is very often useful to do this by comparing the two average heights. Finally, we may not have very many plants available for an experiment, or very much time or labour available to make large numbers of measurements, and so it is important for biologists to study the statistical methods of making comparisons based on small numbers of observations.

I

Populations, Samples and Variates

POPULATIONS

A population may consist of things rather than of people. We have mentioned, in the Introduction, a population of plants; if the height of each plant is measured, the whole set of heights may also be called a population—this time it is a population of measurements.

Before a definition of the word 'population' is attempted, let us consider a few examples.

1 Humans, of specified age-group, racial group and sex may each have their height measured.
2 A field of wheat may be divided into square units ('plots') of fixed size, the crop yield on each plot being weighed.
3 Mice may be bred under controlled conditions in an animal house, and the coat colour of each mouse noted.
4 A properly balanced six-sided die, whose faces are numbered 1, 2, 3, 4, 5, 6, may be thrown a large number of times and the score for each throw recorded.

In (1), the people form a population, and from them we collect a population of height measurements; in (2) the 'plots' form the population (not the individual plants, because we did not weigh each separately) and give us a population of weights; in (3) the animals form the population, from which a population of observations of colour (not, this time, of numbers such as measurements of height or weight) is collected; and in (4) there is only one die, but it is thrown many times to give a population of throws, and hence a population of scores.

We shall use the word 'population' to mean a set of individuals or objects, upon each member of which a numerical measurement is taken or an observation of a particular characteristic made; and we shall also call the set of measurements or observations a population.

Membership of a population must be precisely specified: in (1), we know that, on average, men are taller than women, sons taller than fathers, and some races taller than others, so unless we do have a fairly rigid specification we shall be measuring different ages, races and sexes and collecting a very heterogeneous population of measurements in the process. In (2), the crop records would only be useful if we knew that seed had been sown uniformly and the same variety (or cultivar) of wheat used throughout; in (3) animals should live in the same environment and should be similar genetically (descendants of the same original animals); in (4) the same die should be thrown each time, with adequate shaking to prevent each new result depending on previous ones. Unless we specify a population carefully, we often find some 'unusual' results which, it may be claimed, are not really part of this population; in scientific work we cannot have arguments of this sort casting doubt on the generality of results. When we can say that all measurements in an experiment were taken on members of a carefully specified population, then the experimental results do apply to that population.

Sometimes there are two distinct parts to a population, for example male and female animals, or some plants in a field given additional fertilizer and the remainder not. Then it is useful to take measurements separately on the two parts or *sub-populations*. It is of very little use to make measurements on a very heterogeneous population.

The members actually observed will almost always be a part only, and that often a very small one, of a large population of similar individuals or objects that could have been observed: the observed members form a small *sample* from a larger population. We probably could, with much labour, measure all English males between the ages of 20 and 25 to find their average height; but it would be much more satisfactory if we could measure only a sample, calculate the average height in this sample, and then apply the information from the sample in some way to refer to the whole population. We shall return to this; but quite clearly if the original population has not been precisely defined it will be impossible to produce a sample with any confidence that it represents the original. In some of the examples above, the *complete* original population is infinite and can be imagined only: we could go on breeding mice or throwing dice for ever, so that whenever we stop to look at the results these can be only a sample.

In the social sciences, much use is made of *surveys*, where people are asked questions about such things as their political sympathies or what sort of washing powder they use. There are some notorious pitfalls in obtaining information in this manner, and fortunately the biologist does not often need to work this way. But in a new field of inquiry, the basic problems may have to be discovered by this method: veterinary scientists have had to rely on the observations of farmers and practitioners to find out what may be the urgent problems needing attention, or the conditions which seem to be conducive to particular diseases or disorders.

VARIATES

The measurement (e.g. of height, crop yield, score with a die) or observation (e.g. coat colour of mouse) made on each member of a population or sample is called a *variate* (or sometimes, confusingly, a *variable*). Variates are of two types, **continuous** and **discrete** (or discontinuous). *Continuous* variates are those in which any value whatever (sometimes within certain upper and lower limits) is possible. For example, human height could be 174 cm, 172·72 cm, and even a figure such as 175·0498573 cm if we could measure so accurately: within the known range of human heights, no figure specified, to whatever length in decimals, is impossible. We shall denote the values of continuous variates by x. An example of a *discrete* variate is the score made on throwing a die: it can be only 1, 2, 3, 4, 5 or 6, and any other number is quite impossible—there is no such thing as a face having 1·58 dots on it. Thus we have a limited list of possible numerical results; we denote these by r for a discrete variate. Both of these types may be referred to as **quantitative** (i.e. measurable) variates, whereas the colour of a mouse's coat is an example of a **qualitative** (i.e. observable) variate; these latter variates can frequently be divided into a small number of definite categories, e.g. in mice, white coat, grey coat, brown coat. With sufficient care, quantitative variates can be measured accurately, but qualitative variates lend themselves to subjective errors in their compilation according to which observer actually records them, and so it is extremely important to use a qualitative variate only when there seems to be no equally good quantitative one.

RANDOM SAMPLING

A random sample may be defined for our purposes as one in which every member of the original population has an equal chance of appearing (readers will be able to refine the wording of this definition after the remarks on *probability*). Thus if we have five plants a, b, c, d, e, growing in pots and we decide to choose *at random* two of these for examination in the laboratory, we might equally well find ourselves with a, b; a, c; a, d; a, e; b, c; b, d; b, e; c, d; c, e; or d, e. This would, however, by no means be true if we were to look at the plants and choose two to *represent* the five. Suppose we have labelled them in order of height, *a* being slightly larger than *b*, *b* than *c*, and so on; if we pick *a*, we are most likely to feel that we should take a smallish one, say *d* or *e*, to match it, and we are unlikely to choose *b* deliberately. But if we are going to take a measurement on each of the two sample members, and use the average of these two measurements to make a general statement about what the average for the whole population of five plants might be, the statistical methods of *confidence limits* (Chapter 9) are needed, and these can be used only if the sample consists wholly of members chosen at random. The literature contains plenty of examples of unexpected, subjective errors which have crept in when observers have tried to do their own sampling visually.

In order to choose a random sample from a given (finite) population, it is first necessary to number the members of the population systematically, starting at 1. We next want to obtain a random sample of numbers. Finally we use as our sample the members bearing these numbers. A table of random digits is required, such as that given by Fisher and Yates[7] (Table XXXIII). The aim when producing a table of random digits is that every entry in it is equally likely to be any one of the digits 0, 1, 2, . . ., up to 9, irrespective of entries in any previous position. Random digit tables have been produced by various methods; some simple computer systems have standard programs for generating them when required. Whatever method is used, a set of tests (based on the χ^2 *distribution*, see Exercise 8.11) is made to check whether the digits do appear to be equally likely to come up. If one tries to write down a run of numbers haphazardly out of the head, it is very unlikely that they will conform to this (cf. Exercise 1.3).

Let us suppose that we have a table of random digits, a part of which (the starting-point chosen at random) is as follows:

07435527183494562314358227249611288859777434258800571209 . . . .

This could be used to select 20 members from 1000, which had already been numbered, by grouping the random digits in threes, 074, 355, 271, 834, 945, 623, etc., and choosing as the sample the members of the population carrying these numbers (if 000 appeared, this would be number 1000).

If, for example, number 74 appeared a second time before the full quota of 20 was obtained, there are two choices of procedure:

  Sampling with replacement where the 74th member would actually be used a second time; or *Sampling without replacement*, which is used where the first procedure is impossible for physical reasons, e.g. if an amount of material corresponding to 20 plants was needed for a destructive laboratory test.

Statistical methods are simpler when sampling is with replacement, and we shall assume that this can be done. (However, in the most common situation where the sample forms a very small part, say no more than 5%, of the population, the corrections in the calculations needed *without replacement* are quite negligibly small.)

In everything which follows, we shall assume that populations have been properly specified and samples randomly chosen from them, so that we may concentrate attention on the variates measured or observed.

EXERCISES

(For Answers and Comments, see p. 160)

1.1 Consider how to define precisely these populations:

(a) of students, when an inquiry about expenditure on books is being carried out;

(b) of tomato plants grown in greenhouses, when the number of leaves produced during a given time is measured;

(c) of bean foliage in a smallholding, when numbers of black-fly are counted;

(d) of leaves as in (b) removed for chemical analysis to estimate nitrogen content.

1.2 What type of variate is shown in each part of Exercise 1.1, and in Example (2) at the beginning of the chapter?

1.3 Write down haphazardly a run of 200 digits. Count the number of 0's, 1's, . . ., 9's. Count also the number of times 0 is followed by 0, or by 1, or by 2, etc., and make similar counts for the digits 1 to 9. (The results can be used later as examples of the χ^2-test.)

1.4 How would you use a table of random numbers to select:

(a) 15 members from a population of 750?

(b) 10 members from a population of 250?

(c) 10 members from a population of 300?

(d) 20 sample units, each 1 ft^2, in a rectangular field 25 ft × 40 ft in size?

1.5 Suppose that, in a survey, the following variates have been recorded for several individual people: (a) age, (b) year of birth, (c) sex, (d) height, (e) weight, (f) colour of hair, (g) colour of eyes, (h) surname, (j) age rank in household (i.e. whether they are oldest, second oldest, . . ., youngest person in the household where they live), (k) whether they hold a driving licence. For each of the variates (a)–(k), state whether it is qualitative or quantitative; and, if quantitative, state whether it is discrete or continuous.

1.6 The following practical methods have been suggested for selecting random samples. Say whether each is likely to be satisfactory, and if you consider that it is not, then attempt to propose a better method.

(a) A list of all adults living in a village is available (in the form of the voters' list for elections). In order to survey a random sample of house-holds in the village, to see what sort of fuel they use for domestic heating, names are picked at random from the voters' list.

(b) The doctor who serves this village has a card-index, arranged by families (one card per family). In order to survey the state of the teeth of the children in the village, a random sample of cards is taken, and if there are children in the family shown on a selected card then a random choice is made from among these.

(c) A horticulturalist wants to take a 1-in-10 sample of the strawberry plants growing in rows in a field. He does not have time to select several random numbers and so he chooses a starting-point at random among the first 10 plants in the first row. After this he takes every tenth plant down the first row, back along the second row, down the third row and so on through the field.

1.7   Using the run of random digits near the foot of p. 4 make a random selection of eight letters (without replacement) from the 26 letters of the English alphabet.

2

Summarizing Observed Measurements

EXAMPLE 2.1

Suppose the heights of a large number of plants growing in a green-house are measured. The results, if expressed to the nearest mm, might read: 39, 47, 34, 60, 52, 45, 37, 51, 58, 49, 61, 28, ... Confronted with, say, 187 such measurements, we are in no position to extract much numerical appreciation from these.

EXAMPLE 2.2

Even when a discrete variate with a very limited number of possible outcomes is examined, the position is hardly better: 200 throws of a die may give the scores 3, 5, 3, 1, 4, 4, 6, 4, 2, 1, 5, 3, 6, ...

Before appreciation can really begin, the mass of raw data must be reduced to a summary: the discrete case Example 2.2 is easy to handle, for it is necessary only to list the possible values of r, namely 1, 2, 3, 4, 5, 6, and to count up how many times each occurred—the *frequency* of each value of r, which we denote by $f_1, f_2, \ldots, f_6$, or by f_r as a general symbol. The result of this counting is best expressed as a *frequency table*, for example:

$r =$	1	2	3	4	5	6	*Total*
$f_r =$	31	37	33	30	35	34	200

It is often possible to spot some interesting properties immediately the table is drawn up. In the present case, each of the six scores should turn up about 33 or 34 times among the 200 if the die is properly balanced (*fair*, as it is usually called), and the table suggests that this may well be true:

departures from this 33 or 34 average are small. The raw data, before summary, could not tell us this, though there are other properties which they could indicate that we have now suppressed, such as whether there tended to be quite long runs of each score (e.g. 2, 2, 2, 2, 1, 4, 6, 6, 6, 6, 6, 3, 5, 2, 5, 1, 4, 1, 4, 1, 4, 3, . . .) which could cast doubt on the throwing process even though each score occurs about equally often in the complete set. In this example, it is quite likely that the table would give all the information required, but even so the **cumulative frequencies**, F_r, could be instructive too: F_R is the sum of individual frequencies f_r for all values of r which are less than or equal to R, so that $F_1 = f_1$, $F_2 = f_1 + f_2$, $F_3 = f_1 + f_2 + f_3$, etc., to give:

$$r = \quad 1 \quad 2 \quad 3 \quad 4 \quad 5 \quad 6$$
$$F_r = \ 31 \quad 68 \quad 101 \quad 131 \quad 166 \quad 200$$

This helps to show that there are (about) as many 1's + 2's + 3's (101) as 4's + 5's + 6's (99, i.e. 200–101), so that there is no tendency to show lower scores more (or less) often.

EXAMPLE 2.3

Consider a slightly less simple example of a discrete variate, where the value of r is the number of radioactive particles emitted in a unit of time (say $\frac{1}{2}$ min) from a specimen of plant material which has been grown in a solution containing a labelled nutrient. Suppose that 370 separate $\frac{1}{2}$-min periods were observed, and that when summarized the counts gave this frequency table (to which a third row, showing cumulative frequencies, is added):

$r =$	0	1	2	3	4	5	6	7	8
$f_r =$	2	21	37	50	79	66	49	36	19
$F_r =$	2	23	60	110	189	255	304	340	359

$r =$	9	10	11	12	13	14	15	16+
$f_r =$	7	2	0	0	1	0	1	0
$F_r =$	366	368	368	368	369	369	370	370

Thus r represents the number of particles emitted in one $\frac{1}{2}$-min, and f_r the number of times this particular value of r was counted during the whole period of recording. The pattern in f is less readily seen than in the previous example, and a diagram of some sort would clarify it. A **bar-chart** is the best form of diagram for discrete variates: r is taken along the horizontal (x-axis) and frequency vertical (the y-axis of the graph), and a bar of height f_r is raised vertically at the point r, as in Fig. 2.1. This is quite informative, for we see that there is a pattern of steady increase in frequency as r moves up to 4, followed by another steady (but slightly slower) fall in frequency as r continues through the

Fig. 2.1   Bar-chart showing the frequencies of the numbers, r, of radioactive particles emitted per $\frac{1}{2}$ min (Example 2.3).

values 5 to 10, with very rare r values greater than 10. All this information was, of course, present in the table, but in a less digestible or readily apparent form.

Continuous variates require a somewhat different approach. We cannot list individual values of x and count up frequencies, because unless our sample is very large indeed or the possible range of x-values extremely limited, no single value of x is likely to occur more than two or three times, that is, no single f is likely to be more than 2 or 3. For any helpful summary, the range of x-values has to be split into a number of intervals or **classes**, and the frequency in each of these classes counted up. Suppose that the heights of the plants in Example 2.1 ranged from a lowest value of 26 mm to a highest value of 65 mm, and let us collect together the values in classes 26–30, 31–35, 36–40, ..., 61–65, so that a table of frequencies in each class may be presented as follows:

Variate-value, x mm	26–30	31–35	36–40	41–45	46–50	51–55	56–60	61–65	Total
No. in each class, f_x	4	5	23	58	61	30	3	3	187

Before we can display these results pictorially, or indeed make a numerical summary as in Chapter 3, we must decide exactly what our classification implies. Here, if measurements were to the nearest mm, a plant of height $25\frac{1}{2}$ mm would have been recorded as 26, while one of

$30\frac{1}{2}$ mm would have been called 31: thus category 26–30 contains plants of $25\frac{1}{2}$ mm and all up to, but not including, $30\frac{1}{2}$ mm, so that the *centre point* of this class is 28 mm to the limit of accuracy in measurement used in this case. We shall say also that the ***class-interval*** is 5 mm (the distance between $25\frac{1}{2}$ and $30\frac{1}{2}$, etc.); in the above example *all* the intervals were of width 5 mm and so we say that the above summary is in *equal class-intervals*. But class-intervals are not necessarily always equal, and we might have grouped the first two classes above into one, to contain a frequency of 9, and the last two, to contain 6. A working rule is to have the number and width of classes such that the total frequency is spread over a reasonable number of classes (otherwise we lose a great deal of information in summarizing), but there are not too many classes containing very small frequencies (otherwise we have hardly summarized at all).

For continuous data, the ***histogram*** is the standard type of diagram. In it, the variate-values x are plotted horizontally, and the x-axis is marked off to correspond to the class-intervals chosen. Upon each class-interval as base there is erected a rectangle whose *area* represents the total frequency in that interval: with equal class-intervals, of course, this just means that the height of the rectangle is proportional to frequency. (But if one of the class-intervals, for example, were *twice* the width of the others, the rectangle erected on it would have its height *one-half* the total frequency contained: a little reflection, and re-drawing our example with the first two and last two classes combined, will make this clear.)

Fig. 2.2   Histogram showing the frequencies of occurrence of the plant heights, x (Example 2.1).

The Exercises which follow are designed to present some of the more common shapes of pattern, which will shortly be identified with useful standard distributions.

EXERCISES

(For Answers and Comments, see p. 161)

2.1 The number, r, of males in litters of rats which contain 5 rats altogether, was observed with the following results over 100 different litters. Summarize these results and illustrate them: $r = 2$, 5, 3, 1, 3, 4, 2, 2, 0, 3, 2, 4, 1, 3, 3, 2, 2, 2, 1, 2, 3, 4, 3, 0, 4, 5, 3, 3, 3, 2, 3, 2, 2, 1, 1, 4, 1, 3, 2, 2, 4, 3, 4, 3, 2, 3, 2, 2, 3, 4, 2, 3, 0, 1, 1, 3, 5, 2, 2, 4, 3, 2, 5, 4, 3, 2, 4, 3, 3, 4, 2, 1, 3, 2, 1, 2, 4, 3, 1, 3, 3, 4, 2, 3, 1, 2, 3, 4, 3, 2, 2, 4, 3, 3, 4, 1, 2, 3, 2, 3.

2.2 If the individual frequencies of the values of height in mm recorded for the 187 plants mentioned in the text were as follows, examine the effect of using class-intervals of widths other than 5 mm in summarizing and illustrating the results.

$x =$	26	27	28	29	30	31	32	33	34	35	36	37	38	39
$f_x =$	1	0	2	1	0	1	3	0	1	0	4	1	2	6

$x =$	40	41	42	43	44	45	46	47	48	49	50	51	52	53
$f_x =$	10	12	8	6	15	17	20	13	9	12	7	8	8	6

$x =$	54	55	56	57	58	59	60	61	62	63	64	65
$f_x =$	4	4	0	1	1	0	1	1	1	1	0	0

2.3 The weights (in g) of 120 animals, of similar age and genetic history, at the end of a spell of feeding on the same diet in controlled conditions, were:

Weight $x =$	50–80	80–90	90–100	100–110	110–120
Frequency $f =$	3	6	13	25	24

Weight $x =$	120–130	130–150	150–180	180–240
Frequency $f =$	21	18	7	3

Prepare a suitable diagram to illustrate these observations.

2.4 Why is a histogram inappropriate for discrete data?

2.5 Calculate the cumulative frequencies, F, for the heights of 187 plants as summarized on p. 9 (using 5 mm class-intervals).

2.6 A standard pack of playing cards contains 52 cards in all, 13 of each of the four suits Hearts, Diamonds, Spades and Clubs. A hand of cards consists of 13 out of the 52, and before dealing out a hand the pack must be well shuffled. One hundred hands are obtained in this way from different packs, and the number of Spades in each hand is recorded. Two of these hands contain no Spades, there are 6 hands with 1 Spade, 20 hands with 2 Spades, 28 hands with 3, 23 with 4, 12 with 5, 4 with 6, 2 with 7, and one each with 8, 10 and 11 Spades respectively.

Summarize this information in a suitable table and in a suitable diagram.

3

Distributions and their Characteristic Properties

In Example 2.1 we mentioned a continuous variate, the height of a plant growing in a greenhouse, and gave 187 records of individual heights in Exercise 2.2 (page 11). The table on page 9 showed these records collected into classes of width 5 mm. The first step in making a numerical summary of them is to write down a frequency table whose first row contains the centre points of the classes:

$$x = 28 \quad 33 \quad 38 \quad 43 \quad 48 \quad 53 \quad 58 \quad 63$$
$$f = 4 \quad 5 \quad 23 \quad 58 \quad 61 \quad 30 \quad 3 \quad 3$$

At first sight, this table may give the impression that we are dealing with a discrete variate; but we are now using the frequency 61, for example, to represent the frequency in *the whole class whose centre is* 48 mm, and it is no longer the frequency *at* 48 mm. (That, as we saw in Exercise 2.2, was 9.) We have thus begun to summarize the records by first taking all the frequency in an interval and concentrating it at the centre of that interval. Next we may plot the f-values, in the y-direction, on a graph against the x-values that correspond to each f, as shown in Fig. 3.1.

Joining the plotted points by straight lines gives a ***frequency polygon***, as shown in Fig. 3.1, but a figure with jagged edges does not seem a completely satisfactory way of showing a continuous variate, which ought to vary in a smooth manner. Nor can the mathematical statistician develop any simple methods for us to use on continuous variates unless we assume that there is a *smooth* curve relating f and x. We therefore approximate the

Fig. 3.1 Frequency polygon (straight lines) and frequency curve (dotted) for the data of Example 2.1 as summarized in the frequency table.

frequency polygon with the *frequency curve* shown dotted in Fig. 3.1. Drawing the smooth frequency curve often presents a little difficulty in deciding exactly how to smooth out the sharp corners of the frequency polygon: in this example the behaviour at the peak is not clear (and a slightly different curve would result if the width of class-intervals were changed). In developing mathematical methods, we need an equation for the curve relating f to x, and this is often called a *frequency distribution function*: we shall use the alternative name **probability density function**. In applications, we will not of course need to think very mathematically about a distribution: for us, it is much more important to think of it as the

Fig. 3.2 Frequency curves which are **(a)** symmetrical, **(b)** skew to the right (positively skew), **(c)** skew to the left (negatively skew).

pattern which is followed by f as we go through the x-values in increasing order—the way in which frequency is distributed over the x's.

The main characteristics of shape in a distribution are best seen by looking at continuous variates, though the numerical ways in which these characteristics are measured will be applied to discrete variates also.

1 *Skewness:* distributions may be skew or symmetrical, illustrated by Fig. 3.2(a) which is a symmetrical curve, (b) which is skew with a long tail on the right (positively skew), (c) which is skew with a long tail on the left (negatively skew).

2 *Location:* two curves may be of the same shape, but occupying different positions on the x-axis (because the measurements of x for the set of objects represented by one curve are larger than those for the other); compare (α) and (β) in Fig. 3.3.

Fig. 3.3 (α) and (β): two symmetrical frequency distribution curves having the same variance but different means; (β) and (γ): two symmetrical curves having the same mean but different variances.

3 *Dispersion or scatter:* two curves may again be of the same general shape, but one may contain x-values extending over a much wider range than the other; compare (β) and (γ) in Fig. 3.3.

As we shall see, two or more distributions are usually compared by calculating **measures of location and dispersion** for them; but before doing any numerical comparison at all it is essential to see if the distributions have the same general shape. There is little that could usefully be said in comparing (a) and (c) in Fig. 3.2, for example; it would be best to summarize each one separately. But all too often routine calculations of location and dispersion are made without checking that the distributions involved are of comparable shape: in these days of computers the old advice to *look at the data* needs repeating more than ever and is ignored far too often. We shall content ourselves by making sure that sets of data being compared have all been produced under similar conditions, for example on similar groups of plants or animals in similar environments, where the factor which may be different (say fertilizer or diet) is not so radically different as to affect the shape of the distribution of x. Chapter 16 refers to this kind of difficulty again; meanwhile we note that if the fertilizer or diet given to one group had lacked an essential nutrient element, the growth mechanism of members of that group could very well be upset, so that some care is obviously necessary when assessing the results of experiments like this, where we may have distributions of different shapes to compare.

MEASURES OF LOCATION

Such measures give a general idea of the position of the middle of a distribution. The most common (as well as the oldest) of these is the *average* value of r or x, or in mathematical words the *arithmetic mean*, commonly called simply the **mean**, of all the measurements available. This is denoted by $\bar{r}$ or $\bar{x}$ (r-bar, x-bar), and we have

$$Mean = \frac{Total\ value\ of\ all\ measurements}{Total\ number\ of\ measurements}.$$

Given, as in Exercise 2.1 (p. 11), an unsummarized set of measurements, we simply add them all up and divide by the total number, to obtain (with the aid of a desk or pocket caclulator) $\bar{r} = 260 \div 100 = 2.6$. This is the mean, or average, number of male rats in each litter; to obtain it we have divided the total of male rats for all the litters, 260, by the number of litters, 100. (If it seems odd that this mean is not a whole number—we cannot imagine 0.6 of a rat—remember that it is only a *summary*; a cricketer's batting average is similarly quoted to two places of decimals, though the actual number of runs scored in each single innings must be a whole number.) The summary table of frequencies for this example (see Answers to Exercises, p. 161) provides a much quicker way of finding $\bar{r}$, because it tells us that this particular set of measurements consisted of three o's, thirteen 1's, thirty 2's and so on, that is there were three litters which contained no males each, thirteen litters containing one each, etc. Hence the total of all the measurements, that is the total number of male rats, is $(3 \times 0) + (13 \times 1) + (30 \times 2) + (33 \times 3) + (17 \times 4) + (4 \times 5)$, which of course is 260. Further, the total number of measurements, that is the total number of litters (total frequency), is $3 + 13 + 30 + 33 + 17 + 4 = 100$.

So the mean is
$$\frac{(3 \times 0)+(13 \times 1)+(30 \times 2)+(33 \times 3)+(17 \times 4)+(4 \times 5)}{3+13+30+33+17+4}.$$

We may write this as a formula for general use; let r_i denote a typical one of the set of values that the variate r can take, and let f_i be the frequency of r_i. The symbol $\sum$ (Greek capital sigma) stands for the sum of all expressions like the typical one written down, and is very useful for shortening formulae. If the full list of values of r is $r_1, r_2, r_3, \ldots, r_n$, with corresponding frequencies $f_1, f_2, f_3, \ldots, f_n$, the sum of all the frequencies is $\sum_{i=1}^{n} f_i$, often written $\sum_{i=1}^{n} f_i$: we have written down the typical frequency f_i, we have shown on the $\sum$ symbol the lower and upper limits that i takes in our set of frequencies, and $\sum$ itself just means 'add all these f_i together'. Likewise the total value of all our measurements is $f_1 r_1 + f_2 r_2 + \ldots + f_n r_n$: the typical one of these is $f_i r_i$ and so we write $\sum_{i=1}^{n} f_i r_i$ for the total. (As usual we have used $f_i r_i$ to mean $f_i \times r_i$: multiplication signs are left out in algebraic expressions.) The mean is thus given by the formula

$$\bar{r} = \frac{\sum_{i=1}^{n} f_i r_i}{\sum_{i=1}^{n} f_i}$$

When this is applied to our example, n is of course 6. We shall often find it convenient to have a single letter to stand for the total number of measurements (the total frequency) in a set, and N will be used. So $N = \sum_{i=1}^{n} f_i$.

For a continuous variate like that of Example 2.1, as summarized at the beginning of this chapter, we write x_i for the centre-point of a class-interval, and work out $\bar{x}$ applying the same formula as we have just used for $\bar{r}$:

$$\bar{x} = \frac{\sum f_i x_i}{\sum f_i}$$

For Example 2.1, this gives

$$\frac{\begin{array}{c}(4 \times 28)+(5 \times 33)+(23 \times 38)+(58 \times 43) \\ +(61 \times 48)+(30 \times 53)+(3 \times 58)+(3 \times 63)\end{array}}{4+5+23+58+61+30+3+3}$$

which is $8526/187 = 45 \cdot 59$ mm. We have, of course, lost a small amount of information here by using the summary table, because in the calculation of $\bar{x}$ we have assumed that $x = 28$ occurred four times, $x = 33$ five times and so on. If we had the full details of the original data, as in Exercise 2.2 (p. 11), we would be able to add up all the 187 observations and find that their exact total was 8514. So we could obtain the accurate value of $\bar{x}$, namely $8514/187 = 45 \cdot 53$. This is slightly different from the one calculated by using the summary table, the extent of the difference depending on the class-intervals chosen for the summary and the irregularity of the original data. With discrete variates, no such loss of information occurs.

The value so obtained for $\bar{r}$ or $\bar{x}$ is a summary, in a very concise form indeed, of the full set of original observations. The mean is used a great deal; but sometimes other summary measures are quoted. Two other measures are the mode and the median. The **mode** (or *modal value*) is that value of the variate, r or x, which occurs with the greatest frequency; in the example of a discrete variate above (the data of Exercise 2.1), $r = 3$ is the mode, because there are more 3's than anything else. From the summary on p. 12 of our example of a continuous variate, the largest frequency is 61, applying to $x = 48$ which is thus the mode. Again in Exercise 2.2, the full table of data gives $x = 46$ as the mode, the maximum frequency being $f = 20$; but it is also clear that the unsummarized data, when drawn on a graph, will exhibit several peaks (the frequency at 41 is larger than those for $x = 40$ or 42; 49 gives another local maximum, and so on). It is a fundamental idea in calculating summaries of bio-logical data that the curve which fits the observations really has only one peak, i.e. is *unimodal*. If, even after summarizing the data in a suitable table, it is clear that there are two (or more) considerable peaks, i.e. the data are *multimodal*, no further numerical work should be done until it has been decided whether the observations can be split into two or more sub-populations. We thus take the appearance of two or more modes as a warning that a population may be heterogeneous, and need splitting down into smaller sections before it is worth making calculations of means or any other summary measures. In biological work we hardly ever calcu-late a mode for use as a measure of location.

Before we can calculate the **median**, it is necessary to arrange the list of variate-values in increasing order of size. Then the median, M, is that value of the variate which divides the total frequency so listed exactly into two halves: there is just the same frequency above M as there is below it. To illustrate the method for a discrete variate, consider the data on 370 radioactive counts given in Example 2.3 (p. 8). There are two 0's, twenty-one 1's, and so on; if the full set of 370 is listed in increasing order of size it will therefore begin 0, 0, 1, 1, 1, . . ., with 1's until the 23rd place, then the next 37 places will be occupied by 2's, the next 50 by 3's and continuing in this way to finish . . ., 10, 10, 13, 15. Now M must be such that it has 185 counts above it and 185 below: thus M is midway between the 185th and 186th count in the list. The third row of the table in Example 2.3 is useful here: it gives the cumulative frequencies F_r, and we are looking for values and r_2 of the variate, such that $F_{r_1} = 185$ and $F_{r_2} = 186$; when these are found, $M = \frac{1}{2}(r_1 + r_2)$. There are 110 r-values of 3 or less, and the next 79 are all equal to 4; so $r_1 = r_2 = 4$ and therefore $M = 4$. Whenever the total frequency $N(= \sum_{i=1}^{n} f_i)$ is an even number, we place M midway between the $(\frac{1}{2}N)$th and $(\frac{1}{2}N+1)$th observations; when N is odd the $[(N+1)/2]$th observation itself forms the median, for clearly if N had been 99 the 50th observation would have 49 below it and 49 above.

To calculate M for a continuous variate, we must always place it at the $(\frac{1}{2}N)$th observation, whether N be odd or even; any other rule would not give the same M when calculating from the top as from the bottom. In the 187 plant heights quoted in Example 2.1 (p. 7) we thus seek the $93\frac{1}{2}$th observation. The intervals from 26–30 up to 41–45 contain 90 plants; as we saw on p. 10 and in Exercise 2.5, this means that there are 90 plants whose heights do not exceed $45\frac{1}{2}$ mm. The next interval, beginning at $45\frac{1}{2}$ mm, contains 61 plants; given only the frequency table, in the absence of any more detailed information, we have to assume that these 61 plants have heights uniformly scattered through the interval $45\frac{1}{2}$ to $50\frac{1}{2}$ mm. So we must go $\frac{3\cdot5}{61}$ of the distance through this interval in order to reach the $93\frac{1}{2}$th observation: the complete interval is 5 mm wide, so the required distance is $\dfrac{3\cdot5 \times 5}{61} = 0\cdot287$ mm, and the $93\frac{1}{2}$th observation therefore has $x = 45\cdot5 + 0\cdot287 = 45\cdot79$ mm. So $M = 45\cdot79$ mm.

The median is sometimes preferred to the mean as a summary measure when a distribution is skew or has one or two very extreme (large or small) values; for the inclusion, for example, of one extremely large plant of height 89 mm affects the median hardly at all but increases the mean noticeably. The median is, in this sense, a more stable measure. But the mean is almost universally used in scientific work, because it alone admits of the sort of mathematical background theory which is necessary to apply statistical tests, as described in later chapters.

MEASURES OF DISPERSION

Because the observations making up a set of data are scattered, to a greater or lesser extent, about the mean, the mean alone cannot be a complete summary of them; some measure of this scatter or dispersion is needed also. The common one is the *variance*; this is based on the idea that it is the *size* of the deviations from the mean which is important rather than their direction (above or below). In Fig. 3.3, curves (β) and (γ) are of the same general shape, with the same mean, but (γ) extends over a much greater range of x-values than does (β). If variances were being calculated, that for (γ) would be larger than for (β).

For one particular observation, x_i, its contribution to scatter or dispersion is its distance $(x_i - \bar{x})$ from the mean $\bar{x}$; so as to remove negative signs from the contributions of the smaller x_i, we square to obtain $(x_i - \bar{x})^2$. So if we have a set consisting of N observations, the sum of all contributions to the measure of dispersion is $\sum_{i=1}^{N}(x_i-\bar{x})^2$. The variance is the average of this, denoted by s^2:

$$s^2 = \frac{1}{N-1}\sum_{i=1}^{N}(x_i-\bar{x})^2$$

The divisor is $N-1$, not N, for which a reason cannot be given satisfactorily yet (see Chapters 7 and 8), but a rough-and-ready explanation is that there are N observations in all, and one estimation (of $\bar{x}$) must be made before the remaining $N-1$ units of information can be used in calculating s^2.

The formula presented above makes no use of the fact that a summary table may already be available, especially for a discrete variate, which lists each of the possible variate-values r_i and their corresponding frequencies f_i, so that the total frequency $N=\sum_{i=1}^{n}f_i$ may be dealt with in large parts rather than item-by-item. The only change in the formula is that each $(r_i-\bar{r})^2$ will arise f_i times, so that

$$s^2 = \frac{1}{[(\sum_{i=1}^{n}f_i)-1]}\sum_{i=1}^{n}f_i(r_i-\bar{r})^2$$

This is illustrated using the set of measurements in Exercise 2.1, previously considered on p. 15 where we found their mean to be 2·6. Rewriting the table:

$r_i =$	0	1	2	3	4	5	Total
$(r_i - \bar{r}) =$	−2·6	−1·6	−0·6	+0·4	+1·4	+2·4	—
$f_i =$	3	13	30	33	17	4	100

we can now calculate

$$\frac{1}{(\sum_{i=1}^{n}f_i)-1} = \frac{1}{N-1} = \frac{1}{99}$$

and

$$\sum_{i=1}^{n}f_i(r_i-\bar{r})^2$$
$$= 3(2·6)^2+13(1·6)^2+30(0·6)^2+33(0·4)^2+17(1·4)^2+4(2·4)^2$$
$$= 3\times 6·76+13\times 2·56+30\times 0·36+33\times 0·16+17\times 1·96+4\times 5·76$$
$$= 126·00$$

so that the variance is

$$\frac{126·00}{99} = 1·2727$$

This is not the best method of carrying out this calculation, however: we can use an algebraic fact, which will not be proved here, that

$$\sum_{i=1}^{n} f_i(r_i - \bar{r})^2 = \sum_{i=1}^{n} f_i r_i^2 - \frac{(\sum_{i=1}^{n} f_i r_i)^2}{\sum_{i=1}^{n} f_i} \quad \text{or, even more conveniently,}$$

$$= \frac{1}{\sum_{i=1}^{n} f_i} \left[\left(\sum_{i=1}^{n} f_i \right) \left(\sum_{i=1}^{n} f_i r_i^2 \right) - \left(\sum_{i=1}^{n} f_i r_i \right)^2 \right]$$

The formula looks simpler if we remember that $\sum_{i=1}^{n} f_i = N$, and if we introduce a notation for the total value of all the observations, $G = \sum_{i=1}^{n} f_i r_i$ (G for *grand total*). Then

$$\sum_{i=1}^{n} f_i(r_i - \bar{r})^2 = \frac{1}{N} \left[N \sum_{i=1}^{n} f_i r_i^2 - G^2 \right]$$

so that the variance is

$$\frac{1}{N(N-1)} \left[N \sum_{i=1}^{n} f_i r_i^2 - G^2 \right]$$

For the present example, $G = 260$ was worked out in order to calculate the mean. We need $\sum_{i=1}^{n} f_i r_i^2$; this is

$$3 \times 0^2 + 13 \times 1^2 + 30 \times 2^2 + 33 \times 3^2 + 17 \times 4^2 + 4 \times 5^2 = 802$$

The formula for variance is

$$\frac{1}{100 \times 99} \left[100 \times 802 - (260)^2 \right] = \frac{1}{9900} (80200 - 67600) = \frac{12600}{9900} = 1 \cdot 2727$$

as before. The above method uses the exact values of r_i, and does not require $(r_i - \bar{r})$ to be calculated: this is a great advantage where $\bar{r}$ does not work out exactly to a small number of decimal places as it did in the example, for rounding errors in $(r_i - \bar{r})$ can have a considerable effect on the value calculated for the variance, particularly when this value is numerically small.

When the data simply consist of N distinct observations x_i, the formula

$$s^2 = \frac{1}{N-1} \left[\sum_{i=1}^{N} (x_i - \bar{x})^2 \right]$$

on p. 19 becomes

$$s^2 = \frac{1}{N(N-1)} \left[N \sum_{i=1}^{N} x_i^2 - G^2 \right]$$

which is precisely as above with every $f_i = 1$. This is easy to compute on a calculator, for we simply square every individual x_i, cumulating these squares as we go, then multiply $\sum_{i=1}^{N} x_i^2$, the total of the squares, by the total number N and subtract the square of the grand total G. All these operations can be done in sequence on the machine without writing any intermediate answers on paper, so removing a common source of mistakes.

The final step is to divide by $N(N-1)$, which should have been worked out first and written down. Modern pocket calculators often have built-in programs to calculate the mean and variance, and all we have to do is to feed the data in item-by-item. (This may not be a very good idea for a discrete variate when a summary table is available—it is much quicker to use the summary and the method above.)

Variance, it will be seen, is measured as the sum of several squares, and so is not in the same units as x or r, but is like their square x^2 or r^2: this is a disadvantage for some purposes, and the positive square root of variance, the **standard deviation**, is often useful. For the example above, the standard deviation $s = +\sqrt{1 \cdot 2727} = 1 \cdot 13$.

Other measures of dispersion appear even less often than alternative measures of location: the **range** of a set of measurements is the distance between the smallest and largest numerically, e.g. the data of Exercise 2.2 have a lowest value of 26 mm for x and a highest 63 mm, so that the range is $(63 - 26) = 37$ mm. When only a very few values (not more than six) are available from a normal distribution (Chapter 6), the range is almost as useful as the variance; but in general it is wasteful to use only the two extreme values and ignore the pattern of measurements in between these. The **mean deviation** is $(1/N) \sum_{i=1}^{N} |x_i - \bar{x}|$, where the *modulus* (denoted by $|\ |$) of $x_i - \bar{x}$, namely its numerical value ignoring sign, is employed instead of squaring; it is easier to calculate, perhaps, but is used very rarely. The variance (or standard deviation) is the natural companion of the mean.

EXERCISES

(For Answers and Comments, see p. 162)

3.1 For the data quoted in Example 2.2, on the throwing of dice, calculate the mean, variance and standard deviation.

3.2 For the data quoted in Example 2.3, on radioactive counts, calculate the mean and variance.

3.3 For the data quoted in Exercise 2.3, on the weights of animals, calculate the mean, median, variance and standard deviation; and comment on any difference found between mean and median.

3.4 The following yields (lb) were obtained from plots of fixed size in a field of potatoes growing under the same fertilizer treatment. Calculate the mean, median, variance and standard deviation of these yields.

28, 21, 14, 17, 24, 19, 22, 21, 16, 26, 20, 19, 23, 22, 20, 24, 21, 19,
17, 15, 18, 22, 23, 20, 21, 25, 22, 20, 18, 20, 22, 24, 26, 18, 24, 20.

3.5 The total numbers of organisms of a particular taxonomic group found in each of fifty sampling quadrats were 1, 1, 1, 1, 1, 15, 1, 6, 3, 1, 4, 1, 14, 1, 1, 4, and 0 thirty-four times.   Calculate the mean and variance and comment on the suitability of these as summary measures.

3.6   For the data of Exercise 2.6, on the hands of playing cards, calculate the mean, median and mode, mark these on the diagram which you have drawn, and comment on their similarities and/or differences as measures of location.

Calculate also the variance and standard deviation.

3.7   What are the ranges of the data in Exercises 3.4, 3.5, 3.6? Do you consider that these give any useful information?

3.8   In the data of Exercise 3.4, suppose that two additional plots gave yields of 29 and 30 lb. Examine how the mean and median of the whole data, including these two, compare with those previously calculated in Exercise 3.4, and comment.

3.9   The weights of twelve eggs are 48, 52, 47, 49, 53, 44, 53, 58, 57, 52, 59, 56 g. Calculate the mean and standard deviation of egg weight. We are now told that the first six were laid by hens of one breed, and the second six by hens of another breed. Calculate the mean and standard deviation for each breed, and comment.

4

Probability and the Binomial Distribution

Statistical methods depend on the mathematical theory of probability, and so it is necessary to define this word **probability**; a number of definitions exist, favoured by different schools of thought. The so-called Frequency Theory of Probability is perhaps the easiest to see in practical terms, and is used here in spite of its theoretical limitations.

Consider a very simple experiment, namely tossing a properly balanced coin a great number of times; at each toss, only two outcomes are possible, head (H) and tail (T), and unless some rather unconventional method of tossing is adopted the result of each toss will be unaffected by what has occurred in previous tosses. We may obtain a run of results like THHHTHTHHHHTHHTHTHTTTHHHHTHTT... and this run will often contain more H's than T's, or vice versa: indeed, as the reader may calculate after completing this chapter, it is quite *un*likely that H's and T's will appear exactly equally. At intervals during the experiment, the relative frequency, or *proportional frequency* of heads, p_H, is calculated:

No. of throws n	10	25	50	100	250	500	1000
No. of heads H	7	16	29	54	130	232	481
$p_H = H/n$	0·7	0·64	0·58	0·54	0·52	0·464	0·481

No. of throws n	2000	5000	10000	20000	50000
No. of heads H	1018	2475	5038	9956	25108
$p_H = H/n$	0·509	0·495	0·504	0·498	0·502

This is a typical pattern, in which as n becomes large p_H oscillates about 0·5 with decreasing amplitude: it is reasonable to assume that in the ultimate limit of a very large number of throws indeed (mathematically speaking, as n tends to infinity), p_H will actually be *equal* to $\frac{1}{2}$. We now equate this limiting proportional frequency with the actual *probability* of showing a head in *one single* throw, and say 'the probability of a head is $\frac{1}{2}$'. Another way of thinking about a single throw is to ask ourselves what possible results it could give; leaving out losing the coin, there are only two, H and T. Assuming the coin is properly balanced and is thrown properly, there is no way of claiming that either of H, T is more likely than the other. So half the times we throw, we should show a head, and the other half a tail. Again we say that the probability of a head in one single throw is $\frac{1}{2}$.

A similar thought process goes on when we have only a finite number, N, of members in a population and we know (by having made a count) that m of these are of a specially interesting sort: such as $m = 6$ four-leaved clovers in a patch of lawn containing $N = 84$ clovers altogether. We say that the probability of one clover, picked at random from all these, being four-leaved is $m/N = \frac{6}{84} = \frac{1}{14}$ (one-in-fourteen); here again we are equating *probability* to *proportional frequency*. This rather simple idea of a ratio hardly fits in with the sort of statements made in everyday speech, such as 'It will probably rain tomorrow': for weather-forecasting is so notoriously hazardous that the layman cannot really have enough information to make such a statement other than in a very subjective way. On the other hand, 'My train will probably be late today' may well be firmly based on the knowledge that it has been so every day for a fortnight—or, at the least, more often late than not. (In Chapters 7, 8, 9, a slightly different use of the word is involved.)

If there is reasonable ground for setting up some law about probability (in terms of relative or proportional frequency), we can develop the **Binomial Distribution**. This applies when we divide a population into two parts, so that a proportion p of its members are of a special sort, while the remaining proportion $(1 - p)$ are not. For instance, if we carry out a simple Mendelian plant-breeding trial, p will be $\frac{9}{16}$ for the double-dominant type AA. If we throw a coin, p will (as above) be $\frac{1}{2}$ for heads, and if we throw a six-sided die, $p = \frac{1}{6}$ for any given score (say score 1), that is, we are equally likely to score any of the six possibilities, 1, 2, 3, 4, 5, 6. The binomial distribution finds uses in genetics, but its origin was due to mathematicians' interest in dice-throwing.

Suppose that any single member of the population has probability p of showing a special characteristic (double-dominance; heads; score 1), and thus $(1 - p)$ of not showing it: we denote $1 - p$ by the symbol q. Since every member either does or does not show the characteristic,

$p+q$ *does* have to equal 1: we really can classify every member of the population (no coins landing on their rim, or dice on their corners, when thrown). Remember that p, q are proportions, *not* percentages; this will be found to make formulae tidier, but when figures are given in percentage form, they must be turned into proportions, so that 52% is $p = 0.52$, $7\frac{1}{2}\%$ becomes $p = 0.075$, etc. This definition of p implies that $p = 0$ only when there are no members of the special sort, i.e. it is *impossible* to find one; and $p = 1$ when all members are of the special sort, i.e. it is *certain* that any member examined will be so.

Now take a sample of n members at random from the whole population, independently of each other (so that the choice of one member implies nothing about what other members are chosen); in the genetic example, this means looking at n plants. The number, r, an integer between 0 and n, will denote how many out of this sample of n have the special characteristic under study, which for the plants might be that they show double-dominance; r is a discrete variate, since it can take only the integer values $0, 1, 2, 3, \ldots, n$, not all of which are equally likely. Let us call $P(r)$ the probability that the variate has the value r, e.g. the probability that exactly r of the n plants will be double-dominants; then the complete list of values $P(0), P(1), P(2), \ldots, P(n)$ gives the probability distribution of r. (The remaining $(n-r)$ plants are then not double-dominants, but may be Aa or aa.) This distribution differs from the examples of distributions we have already seen only in that its (proportional) frequencies are calculated by a theoretical argument, as follows, rather than from measurements made on a set of members of a population.

For each chosen member of the sample, the probability is p that it will be of the special type, q that it will not. Moreover, each sample member is chosen independently of every other: this means that the probability of a particular sequence of choices is the product of the probabilities of the individual choices which constitute the sequence. It is easiest to see this in a simple case, namely a sequence of two tosses of a coin, for which we wish to calculate the probability of obtaining two heads. In half of the total number of sequences (if we go on doing this indefinitely), a head comes first, and in half *of these* cases a head also comes second: the phrase *of these* tells us to *multiply* the probabilities, $\frac{1}{2} \times \frac{1}{2} = \frac{1}{4}$. Any of the other possible sequences (head-tail, tail-head, tail-tail) will also have probability $\frac{1}{4}$. There are, in fact, four possible results (HH, HT, TH, TT), all of equal probability, which is another way of seeing the answer.

The sequence in the binomial case contains r special members, each with probability p, and $(n-r)$ not of the special type, each with probability q, leading to a total probability of $p^r \times q^{n-r}$. But there is a further point to note: $P(r)$ is the probability of obtaining r of the one sort and $(n-r)$ of the other, *in some order or other*, the actual order being immaterial. Consider coin-tossing: there are two ways of obtaining one head and one tail, namely HT and TH. So, in general, in order to calculate $P(r)$ correctly, we must allow for the number of ways in which a sample of n may be split so as to contain r and $(n-r)$, and multiply the probability $p^r q^{n-r}$ by this number of ways. Here a new symbol is needed, called n-factorial, written $n!$, and standing for $n \times (n-1) \times (n-2) \times \cdots \times 3 \times 2 \times 1$. The required number of ways of picking r out of n is (without proof)

$$\frac{n!}{r!(n-r)!}, \quad \text{i.e.} \quad \frac{n(n-1)(n-2)\ldots 2.1}{[r(r-1)\ldots 2.1][(n-r)(n-r-1)\ldots 2.1]}$$

This is frequently written ^nC_r, or $\binom{n}{r}$.

By convention, $0!$ is taken as 1, so that

$$\binom{n}{n} = \frac{n!}{n!0!} = 1$$

as might be hoped in common sense. Before using this in the Binomial formula, we first show that $\binom{n}{r}$ *is*, in a few simple cases, equal to the number of ways of choosing n members so as to contain r of one sort (which we call x) and $(n-r)$ of another (call it o). When $n=4$ and $r=2$, we want to see how many ways we can find of writing $2x$'s and $2o$'s: the answer is *xxoo, xoxo, xoox, oxxo, oxox,* and *ooxx,* namely 6. Now

$$n! = 4 \times 3 \times 2 \times 1, \qquad r! = 2 \times 1, \qquad (n-r)! = 2 \times 1$$

so that

$$\frac{n!}{r!(n-r)!} = \frac{4 \times 3 \times 2 \times 1}{2 \times 1 \times 2 \times 1} = 6$$

And when $n=5$, $r=3$, we have to select 3 x's and 2 o's, in any of the 10 ways *xxxoo, xxoxo, xxoox, xoxxo, xoxox, xooxx, oxxxo, oxxox, oxoxx, ooxxx.* Here

$$\frac{n!}{r!(n-r)!} = \frac{5!}{3!2!} = \frac{5 \times 4 \times 3 \times 2 \times 1}{3 \times 2 \times 1 \times 2 \times 1} = 10$$

The Binomial formula now becomes

$$P(r) = \binom{n}{r} p^r q^{n-r} \quad \text{for } r = 0, 1, 2, \ldots, \text{ up to } n \tag{1}$$

The reason for the name *Binomial Distribution* will now be clear to those readers who have met the so-called *binomial series* expansion of $(1+x)^n$; for in (1) the values of $P(r)$ are the successive terms in the expansion of $(q+p)^n$. It also helps to remind us that we can use this distribution only when a population can be split into just *two* parts (*bi-* means *two*): before trying to apply it in practice we look at the problem to make sure there is one type of population member we want to study, and all the rest of the members are of some other type (or types) in which we are not interested. Then p must be the proportion of the interesting type among the whole.

EXAMPLE 4.1

Consider tossing $4(=n)$ coins, which are properly balanced so that the probability of a head at each toss is $p=\frac{1}{2}$. Let r be the number of heads resulting from the 4 tosses; r may be 0, 1, 2, 3 or 4, and the r heads may arise anywhere in the sequence of 4 tosses. So

$$P(r) = \binom{4}{r}\left(\frac{1}{2}\right)^r\left(\frac{1}{2}\right)^{4-r}$$

The probability element $(\frac{1}{2})^r(\frac{1}{2})^{4-r}$ is the same whatever r, namely $(\frac{1}{2})^4$ or $\frac{1}{16}$. When $r=0, 1, 2, 3, 4$ the respective values of $\binom{n}{r}$ are 1, 4, 6, 4, 1. Thus $P(0)=\frac{1}{16}$, $P(1)=\frac{4}{16}=\frac{1}{4}$, $P(2)=\frac{6}{16}=\frac{3}{8}$, $P(3)=\frac{4}{16}=\frac{1}{4}$, $P(4)=\frac{1}{16}$.

This list of probabilities may be interpreted in various ways. Let us call one set of four tosses a *trial* (the word experiment is often used in this context, but we avoid it to prevent confusion with the type of experiment considered later in the book). Then one interpretation of the probabilities is that *on average*, repeating the trial a large number of times, we expect to obtain no heads only once every sixteen trials, one head once every four trials, two heads three times every eight trials, three heads once every four trials and four heads once every sixteen trials. Alternatively we may concentrate our attention on what happens in a *single* trial: clearly we can obtain any one of the five possible results 0, 1, 2, 3, 4 heads, and the list of probabilities shows which results are more, and which less, likely when only one trial is made. Probability can never tell us what *is going* to happen in a single trial; instead it gives us an idea which of the various possible results to *expect* and which to be surprised at.

Note that $\sum_{r=0}^{4} P(r)=1$, since the list $P(0)$, $P(1)$, $P(2)$, $P(3)$, $P(4)$ accounts for all possible results which the four tosses could give, i.e. for a proportion 1 (100%) of possible results. Fig. 4.1a illustrates this set of probabilities.

EXAMPLE 4.2

When $p = \frac{1}{2}$, the probabilities are easy to calculate. But now consider the results of a simple genetic trial, where a $3:1$ segregation of the two types A:a is expected. Thus p, the proportion of A's, is $\frac{3}{4}$. Suppose that $n = 10$ members of the population are examined: r will be the number of A's, which may equal 0, 1, 2, ..., 9, 10. Applying

$$P(r) = \binom{10}{r}\left(\frac{3}{4}\right)^r\left(\frac{1}{4}\right)^{10-r} = \binom{10}{r}3^r\left(\frac{1}{4}\right)^{10}$$

the values are found to be:

$r = 0$	$P(r) = 0.000001$		$r = 6$	$P(r) = 0.145998$
1	0.000029		7	0.250282
2	0.000386		8	0.281568
3	0.003090		9	0.187712
4	0.016222		10	0.056314
5	0.058399			

which are shown in Fig. 4.1b. The trial here consists of examining $n = 10$ members; the result of this trial is to observe the number, r, of A's, and so we see that the result *no A's* will on average occur only once in one million trials. Alternatively we may say that it is extremely unlikely that in one single trial we shall find no A's. The modal value is 8 A's (the largest value of $P(r)$ occurs when $r = 8$). The total probability for $r = 0$, 1, 2, 3, 4, 5 together is 0.078127, by adding the six separate probabilities; this says that, in a single trial, we are quite unlikely to find a number of A's which is 5 *or less*. We are very likely to find 6 or more A's, because the probability of $r = 6$ or more (that is, 6, 7, 8, 9 or 10) adds to 0.921874.

Given a distribution built on a theoretical argument, it is reasonable to ask if its mean and variance can be expressed in any simple way. We define a **parameter** of a distribution as follows: a formula such as (1) describes a whole *family* of distributions, and it is only when specific numerical values are assigned to n and p (q, being $1 - p$, is then determined also) that one particular pattern of $P(r)$ can be calculated as has been done in the examples. Then n, p are parameters whose numerical values fix exactly which one of the binomial family is being employed. So any general expression for mean or variance will have to be an algebraic one containing the parameters. For the binomial distribution in general we find that mean $= np$ and variance $= npq$. In the two examples above, this gives mean $= 4 \times \frac{1}{2} = 2$ and variance $= 4 \times \frac{1}{2} \times \frac{1}{2} = 1$ for Example 4.1; and for Example 4.2, mean $= 10 \times \frac{3}{4} = 7.5$ and variance $= 10 \times \frac{3}{4} \times \frac{1}{4} = 1.875$.

Fig. 4.1   Bar-charts for the values of $P(r)$ in the Binomial distributions (a) with $n = 4, p = \frac{1}{2}$ (Example 4.1); (b) with $n = 10, p = \frac{3}{4}$ (Example 4.2).

In Chapter 6 we shall note the use of mean and variance to summarize a distribution even when the equation of its probability density function is not known exactly; but at present it is clear that for the binomial distribution n and p provide a complete summary if their numerical values are known. In fact, n and p are the important things about a binomial: mean and variance are subsidiary to these.

We now consider another related point. If we are given a sample of data, and have reasonable grounds for expecting them to follow a binomial distribution, how shall we estimate the parameters of this distribution? Consider again the data of Exercise 2.1; 100 sets of samples of $n = 5$ rats were examined, and it seems very reasonable to expect the probability p that any individual rat is a male to be constant throughout the population: in other words, an estimate of p really would have a useful meaning. In this case we know the parameter n; it is 5, the size of the litters examined. (If this size were not constant, the necessary calculation would be much less straightforward than that which follows.) A number, N, of these litters, in this case 100, were examined and gave the information that $\bar{r} = 2.6$. Considering these as a random sample from the whole population of litters of animals being studied, we may thus say that the sample mean is 2·6. An estimate of the true value of p in the population is obtained by making the statement *Sample Mean = Population Mean*. This is a general method which will be applied to other families of distributions also; for the binomial it requires setting the sample mean equal to np, and hence providing an estimate of p. So we obtain $2.6 = 5\hat{p}$, whence $\hat{p} = 0.52$; the symbol $\hat{p}$ is often used to denote an estimate of p that has been calculated from sample data. Hence if the data of Exercise 2.1 really are binomial, the parameter $n = 5$ and an estimate of p is 0·52. (Naturally, if a different set of litters is examined, a slightly different estimate of p may arise, and we return to this in Chapter 9; a further problem, namely how we may check whether the data really are binomial, is dealt with in Chapter 8.)

When there is no reason for p to remain constant from one member, or one part, of a population to another, the binomial distribution will *not* give a satisfactory explanation of observations; for the binomial to be applicable, it is essential that every member sampled shall be subject to the *same* probability or proportion p. We have mentioned earlier how important it is to make sure that the population being studied is *homogeneous* before applying statistical methods to it. When studying genetic characteristics using the binomial, what matters is that every member examined shall have the *same* probability of showing the particular characteristic being studied.

EXERCISES

(For Answers and Comments, see p. 163)

4.1 If the four types AB, Ab, aB, ab, occur in the ratio $9:3:3:1$ in a population, what is the probability that

(a) one randomly selected member will be an AB?
(b) one randomly selected member will be either an AB or an Ab?
(c) when 2 members are selected at random, the first will be aB and the second Ab?
(d) when 2 members are selected at random, one will be AB and the other ab?

4.2 A plant breeder knows by experience that when he crosses 2 varieties of currant, 10% of the offspring will be disease-resistant, and that only 20% of the offspring will be sufficiently vigorous growers to warrant further trial. If he wishes to breed plants which are both vigorous and disease-resistant, and these two characteristics are independent of one another, what proportion of offspring will be of use to him?

What is the probability that, in 100 such offspring, none will be worthy of further study? And what is the mean value of the number of useful offspring in 100?

4.3 Seeds of a certain plant have a 60% germination rate. Calculate the probability that, when 8 of these seeds are planted, 6 or more will germinate.

4.4 Persons suffering from a blood disease are found to have an abnormality in one particular chromosome. However, not all samples of this chromosome are abnormal, and in order to estimate the proportion of affected ones, 5 examples of this chromosome are examined from each of 120 patients, the number of affected ones, r, being recorded for each patient. The results are tabulated below; estimate the proportion p of affected chromosomes.

$r =$	0	1	2	3	4	5	Total
$f =$	6	31	42	29	10	2	120

4.5 Six plants a, b, c, d, e, f are grown from the same source of seed in the same environment, and at the end of the growing season they have reached the following heights: a is 25 cm high, b is 18, c is 32, d is 28, e is 24, f is 21. Two plants are selected at random from these six; that is, every possible pair has the same chance of occurring. How many different samples of two are there? Find the mean height in each possible sample, and show that the mean of these means is equal to the mean height of the original six plants.

What is the probability that plant a is one of the two chosen when a pair is picked at random?

4.6 Explain why you would, or would not, expect the measurements described below to follow a binomial distribution. For those which do, state the values of n and p.

(a) Twelve animals are selected at random from a large population in which a particular type of marking occurs with probability $\frac{1}{3}$ on each member. The number having the marking in the sample is recorded.

(b) A penny is tossed until it first shows a head. The number of tosses required for this is recorded.

(c) A group of 40 students contains 10 who are left-handed. Five students are chosen at random from this group, and the number of left-handed members in these five is recorded.

(d) An animal is being trained to traverse a maze correctly. At each session it is allowed 5 runs at the maze, and the number of correct runs is recorded.

(e) Groups of 4 people each answer a psychological test in the form of a battery of questions, and it is expected that any person has an equal chance of scoring above or below half the total 'marks'. The number in each group scoring above half is recorded.

4.7 In a food production process, packaged items are sampled as they come off a production line: a random sample of 5 items from each production batch is checked to see if each is tightly packed. A batch is accepted if all 5 sample items are satisfactory, and rejected if there are 3 or more unsatisfactory packages in it; otherwise a further sample is taken before making a decision. If in fact the packing machine is giving 80% of items properly packed, what is the probability that this second sample will be necessary?

The second sample also consists of 5 items. What is the probability that, out of the two samples (10 items), there are 9 satisfactory?

What are the assumptions behind your calculations?

5

The Poisson Distribution

The Poisson distribution takes its name from the French mathematician who first studied it. He was interested in a population which contained only a very small proportion of members with a given characteristic (that is to say, p of the previous chapter is very small), but a large sample could be taken (that is, n is large). If n is large, a very tedious amount of calculation of binomial probabilities would be needed, and Poisson's distribution is an approximation to the binomial in which $P(r)$ is much easier to calculate. Mathematically, he let p become small at the same rate as n became large ($p \to 0$, $n \to \infty$) and so made the product np remain finite: we shall write λ for this product ($\lambda = np$) and from now on think in terms of λ instead of n and p separately. We are still counting distinct items, so we still have a discrete distribution; r is the number of our rare type in a very large sample, and therefore takes the values $r = 0, 1, 2, 3, \ldots$ without specified upper limit. The formula (**11**) (p. 34) for $P(r)$ contains only one parameter, λ; the Poisson distribution is specified by this *single parameter*.

Although for many years the Poisson distribution was considered as only an approximation to the binomial, it has recently been shown to apply more directly when the conditions (*i*), (*ii*), (*iii*) below are satisfied. These lead to a much better appreciation of what the distribution is, and we give some illustrations before stating the general conditions.

First, suppose that we are counting blood cells on a slide under a microscope, that the suspension containing them is well diluted to separate them out and that the area of field studied is large in relation to the size of each cell. Denote by r the number of cells counted on a slide; then as we continue counting many slides, the number r will follow the Poisson distribution.

Secondly, consider counting small insects on the leaves of a host plant, where the insects do not tend to group together in clumps nor do they show any preference for upper, lower, older or younger leaves. If r denotes the number of insects per unit area of leaf, and counting is extended over several unit areas of the leaves of several plants, r will follow the Poisson distribution.

The general conditions under which the distribution arises are:

(i) r is found by counting the number of individual items of the same type (e.g. cells or insects) which are present within a given unit area;

(ii) each individual item occurs purely *at random* within the area;

(iii) each individual item occurs *independently* of every other item.

The second condition implies that any particular spot in the area is equally likely to carry a cell or an insect: there is no tendency for cells to drift to the edges of slides or for insects to prefer some specific parts of the foliage. Condition (iii) requires that the presence of a cell at a particular point on a slide has no influence on whether or not there is another cell close by it—cells neither attract nor repel one another; similarly the presence of an insect on a leaf has no bearing on whether or not another insect will choose the same portion of leaf.

A further example of the Poisson distribution is the count of the numbers of plants of a given species which are found in sampling quadrats of a suitable size, if the plants have arisen from seeds distributed at random over the area sampled. Here it is possible for condition (iii) to be violated if the quadrats are too small relative to the size of the plant. In ecology, the Poisson distribution is often taken as the standard 'random' distribution of species over an area.

Other examples of this distribution arise when counts are made over suitable units of time rather than areas of space: the number of telephone calls going through an exchange in, say, 5-min periods (assuming callers to act independently of one another and effectively at random in time), or emissions from a source of radioactive particles provided these are independent and random. When events in time are considered, condition (i) requires us to use fixed units of time (which may be anything that is convenient, e.g. 1 min, $\frac{1}{2}$ min, $\frac{1}{4}$ hr or longer if the individual events do not occur very often), condition (ii) implies that the telephone calls, radioactive emissions, etc. may happen at any instant of time, and (iii) does not allow, for example, either dependent chains of emissions of particles or a dead period after each emission.

The conditions (i), (ii), (iii), satisfied by the Poisson distribution are the simplest ones for the development of a so-called *stochastic*, or random, process: the total number of telephone calls which have passed through by the end of, say, an hour has been built up in a random manner; so has the total number of cells counted by the time a field has been completely scanned. Stochastic processes have been much studied in recent years, not least in biology, and the Poisson distribution has taken on a new importance in no way related to its original derivation.

The formula for $P(r)$ contains the mathematical constant e, which some readers will know as the base of natural logarithms (i.e. $\log_e e = 1$)*; for computation it is necessary only to know its numerical value $2 \cdot 7183$ (so that e^2 means $(2 \cdot 7183)^2$, which is equal to $7 \cdot 389$). There is also the expression $e^{-\lambda}$; this means $1/e^\lambda$, and hence e^{-2} means $1/e^2$, i.e. $1/7 \cdot 389$ which equals $0 \cdot 1353$. But λ may not be a whole number; in this case the expression $e^{-\lambda}$ can easily be calculated using logarithms, so that if $\lambda = 1 \cdot 6732$, $e^{-\lambda}$ is $1/(2 \cdot 7183)^{1 \cdot 6732}$; the log of this is $\log 1 - \log(2 \cdot 7183)^{1 \cdot 6732}$ which is $0 - 1 \cdot 6732 \times \log 2 \cdot 7183$. There is no need to have a table of natural logarithms for this calculation; logs and antilogs to base 10 may be used to obtain $\log_{10}(e^{-1 \cdot 6732}) = -1 \cdot 6732 \times 0 \cdot 4343 = -0 \cdot 7267 = \overline{1} \cdot 2733$, and taking antilogs gives $e^{-1 \cdot 6732} = 0 \cdot 1876$. In fact, anyone with a pocket calculator will be able to compute e^x, whatever x is, simply by using the appropriate key.

We quote without proof the formula giving $P(r)$ when r follows the Poisson distribution

$$P(r) = \frac{e^{-\lambda}\lambda^r}{r!} \quad \text{for } r = 0, 1, 2, 3, \dots \tag{11}$$

When $r = 0$, we shall obtain simply $P(r) = e^{-\lambda}$, because $0!$ is equal to 1 and also $\lambda^0 = 1$ whatever value λ may take.

Theoretical study of this distribution reveals that *its mean and variance are both equal to* λ. Thus λ gives the average number of telephone calls through an exchange in unit time, or radioactive particles emitted in unit time, the average number of blood cells per unit area of microscope slide, or insects per unit area of leaf, or plants per unit area within sampling quadrats.

EXAMPLE 5.1

Suppose we count the number, r, of a special type of cell present in unit area of a suspension on a microscope slide, and find that, on average, there is one of these cells in every 2 units of area, i.e. $\lambda = \frac{1}{2}$. If the counts r follow a Poisson distribution, we expect that

$$P(0) = e^{-\lambda} = e^{-\frac{1}{2}} = \frac{1}{\sqrt{e}} = \frac{1}{\sqrt{2 \cdot 7183}} = 0 \cdot 6068$$

* A guide to the use of logarithms is given in companion volumes of the 'Contemporary Biology' series. These are *A Biologist's Mathematics*, by D. R. Causton, Chapter 2, Numbers, Indices and Logarithms, and *A Biologist's Physical Chemistry*, by J. G. Morris, Chapter 1, Mathematics Revision.

$$P(1) = \frac{e^{-\lambda}\lambda^1}{1!} = \frac{1}{2} \times e^{-\frac{1}{2}} = 0.3034$$

$$P(2) = \frac{e^{-\lambda}\lambda^2}{2!} = \frac{e^{-\frac{1}{2}}(\frac{1}{2})^2}{2 \times 1} = 0.0759$$

similarly $P(3) = 0.0126$; $P(r \geq 4) = 0.0013$. The sum of all the probabilities $P(r)$, taken over all possible values of $r = 0, 1, 2, 3, \ldots$, to infinity, must equal 1; in fact the first few values $P(0)$, $P(1)$, $P(2)$, $P(3)$ add up to almost 1 in this example, and so we express the probability of all the remaining values, $P(r \geq 4)$, in a single number. The Poisson distribution tails off very quickly when λ is small, and quite often it is necessary to work out only relatively few probabilities.

EXAMPLE 5.2

Example 2.3 gave data in which r denoted the number of radioactive particles emitted in a suitable unit of time (say $\frac{1}{2}$ min: this would depend on the rate of emission) from a specimen of labelled plant material. The data were summarized in Chapter 2 and shown in Fig. 2.1. This is a typical pattern for the Poisson distribution when λ is greater than 1, the modal value being reached quickly (here it is at $r = 4$) and the graph tailing off to the right more slowly. The mean of this set of data is found to be 4.59 and the variance 4.41.

An estimate of λ is obtained by equating the sample mean to the theoretical mean (cf. p. 29 where this method was used for the binomial distribution). For Example 5.2 this gives $\hat{\lambda} = 4.59$, since the theoretical value of the mean is λ and the actual value of the sample mean is 4.59. The sample variance is almost equal to the sample mean for this set of data, which thus appear to satisfy the characteristic property of the Poisson distribution, *mean = variance*. For a proper test of whether the data really do fit the Poisson distribution, see Chapter 8 (p. 66).

We have just noted that for the Poisson distribution mean = variance; sometimes a set of data show variance > mean, and then we suspect that independence of individuals is no longer true, but that instead there is a clumping effect: the presence of one makes the presence of others more likely. An example of this is found when counting insects' eggs, which will originally have been laid in batches rather than individually, though it is quite likely that the batches will have been laid at random. However, batches will not be of constant size, and in order to allow for this it is necessary to superimpose a distribution describing the variation in batch size on to a Poisson distribution for the batches themselves. We refer readers to more advanced texts for the detail of this calculation; the result is a distribution known as *negative binomial* (because its successive terms can be found mathematically using the expansion in series of $(1 + x)^{-n}$). It has found various uses in biology, generally when conditions similar to the above have arisen, and is another of the distributions often used in ecology. A slightly different case is the study of *accident statistics*, that is, the number of accidents occurring per week in a factory, for example: the negative binomial often fits here, and it has therefore been suggested that accident-*situations* arise at random, but that people vary in accident-*proneness*, i.e. are not all equally likely to have an accident in such a situation. So the Poisson distribution explains the accident-situations rather than the resulting accidents.

EXERCISES

(For Answers and Comments, see p. 164)

5.1 A radioactive source is emitting, on the average, one particle per minute. If counting continues for several hundred minutes, during which time the particles are emitted randomly, in what proportion of these minutes is it to be expected that

(a) there will be no particle emitted?
(b) there will be exactly 1 particle emitted?
(c) there will be 2 or more particles emitted?

5.2 The numbers of snails found in each of 100 sampling quadrats in an area were as follows:

Number of snails, r	0	1	2	3	4	5	8	15
f = frequency of r	69	18	7	2	1	1	1	1

Find the mean and variance of the number of snails per sampling unit, and use these to gain a rapid idea of whether the data fit the Poisson distribution.

5.3 The numbers of plants of a certain species falling in 100 randomly chosen quadrats in an area were counted as follows:

Number of plants	0	1	2	3	4	5	6	7	8	Total
Number of quadrats	17	20	28	18	8	8	0	0	1	100

Does this suggest that the plants are growing at random in the area?

5.4 Random samples, each of unit volume, are taken from a well-mixed cell suspension in which on average one cell of a special type is present per unit volume. Find the probability that a sample will not contain more than one of the special cells.

If ten such samples are taken, find the probability that exactly five of these will contain not more than one of the special cells.

5.5 The average density of a bacterial organism in a liquid is μ per ml. Fifty separate 1 ml units are taken from the liquid, and ten of these prove to be sterile (i.e. contain no organisms). Find an estimate of μ.

6

The Normal Distribution:
its Occurrence

This distribution is of great importance in theoretical as well as applied statistics, and has been studied by various writers; sometimes it is called *Gaussian*, but in using the name *normal* we follow common biological practice. (K. Pearson, the British biometrician, seems to have invented the name 'normal', at the end of last century.) It is a continuous distribution, and so is fitted to the task of explaining the variation in such measurements as the heights of human beings or of plants (Example 2.1). Because it will turn up so often in later chapters, we must not immediately assume that almost anything can safely be treated as normal: there are a number of ways of justifying its use, but it is none the less foolish to use it without seeing that one of these ways is relevant. We return to this later in the chapter.

When the variate being measured is a continuous one, we choose to call it x (rather than r), so that the **mathematical** equation for the frequency of occurrence at a variate-value x will be a *function of x*; in other words it will be an expression whose actual numerical value can be evaluated if a particular value is given to x. The frequency at x is called $f(x)$, and when a graph of $f(x)$, in the vertical direction, against x, in the horizontal direction, is plotted we obtain the familiar bell-shaped curve of Fig. 6.1. We should obtain the same bell-shape if we collected a large amount of data on human heights or plant heights (from a homogeneous population), summarized it and drew a histogram or frequency polygon, and then tried to put a smooth frequency curve on top of this as illustrated in Fig. 3.1. The normal distribution was of course discovered in this way by studying large collections of data.

The equation is

$$f(x) = \frac{1}{\sigma\sqrt{2\pi}}\, e^{-(x-\mu)^2/2\sigma^2}$$

where e is the mathematical constant already encountered in the Poisson distribution and π is the very familiar constant ($3\cdot1416$). The actual

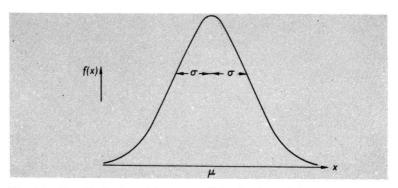

Fig. 6.1   The distribution of frequency, $f(x)$, of values of the variate, x, in the normal distribution with mean μ and variance σ^2.

equation need not concern us greatly, since all the properties of the normal distribution used in applied statistics can be appreciated graphically or with the aid of tables. Only two points need to be noted: first, that in theory there is no upper or lower limit to the possible values of x, though the frequencies become extremely small for values of x some distance from the centre of the distribution. Secondly, the equation contains (besides e and π) the symbols μ and σ^2. These are *parameters* of the distribution, and we have anticipated their meaning in the symbols we have given them, for μ proves to be the *mean* of the distribution and σ^2 its *variance*. (In earlier chapters, we have used $\bar{r}$ or $\bar{x}$ for mean and s^2 for variance: these were *estimates* of mean and variance, based on sample data, and we reserve the Greek μ, σ^2 symbols for the true values of mean and variance in complete populations or—as here—in theoretical distributions.) The normal distribution is symmetrical about μ (which is also the point of maximum frequency) and so is appropriate when a measurement displays a strong tendency towards taking its mean value, from which deviations are equally likely to be in either direction. The standard deviation, σ, measures the distance from μ of the point-of-inflexion in the frequency curve; in simple terms, we think of σ as measuring the width or scatter of the distribution, for in Fig. 3.3 the two curves (β), (γ) are both normal with the same mean, but (γ), being *wider*, will have the larger value of σ.

From what we have said, it will be clear that Fig. 6.1 represents the characteristic shape of the whole *family* of normal curves, and that in order to fix exactly which member of this family we wish to use in any particular case we need to give a numerical value to its mean, μ, and also to its variance, σ². For this reason we shall specify particular curves by the symbol $\mathcal{N}(\mu, \sigma^2)$, quoting within the brackets the numerical values of mean and variance (note that we use σ², not σ). Thus $\mathcal{N}(4, 5)$ and $\mathcal{N}(7, 5)$ are two normal curves of the same width (their variances both being 5), but the second is situated further to the right on a graph of $f(x)$ against x, since it has a greater value for its mean (7) than does the first one (μ = 4). Further, $\mathcal{N}(4, 5)$ and $\mathcal{N}(4, 1)$ both have the same mean, 4, but the second curve is much narrower than the first because its variance is much smaller.

The normal distribution has been extensively studied in the *standard form* $\mathcal{N}(0, 1)$: any $\mathcal{N}(\mu, \sigma^2)$ can be turned into $\mathcal{N}(0, 1)$ as follows. If the variate x is distributed as $\mathcal{N}(\mu, \sigma^2)$, then $(x-\mu)/\sigma$ is $\mathcal{N}(0, 1)$; in other words, if we consider, instead of x, the deviation of x from its mean μ (with care to include a minus sign if x is less than μ), and measure this in units of its standard deviation σ (so using σ as the unit in a new scale of measurement), we have a value which follows the *unit (or standard) normal distribution* whose mean is 0 and variance 1. For example, if we have a value 10·3 from $\mathcal{N}(8\cdot2, 9\cdot0)$, this corresponds to, and can be treated as, a value $(10\cdot3 - 8\cdot2)/3\cdot0 = 0\cdot7$ from a $\mathcal{N}(0, 1)$. In significance testing (Chapters 7, 8) normal distributions are always used in standard form, and tables of the normal distribution give, for particular values of x, the *cumulative probability* that a $\mathcal{N}(0, 1)$ variate will take a value less than or equal to x. The cumulative probability is $\frac{1}{2}$ for $x = 0$, because this value of x is the middle of the distribution, and the total probability for the whole curve has been made 1 by putting in the constant multiplier $1/(\sigma\sqrt{2\pi})$. In Chapter 7 the use of such cumulative-probability tables in significance testing will be explained.

The normal distribution occurs naturally, as we have already seen, in summarizing data on the heights of human beings, or of plants, or in the well-known agricultural example of the crop yields from individual *plots* (i.e. small areas) in a wheat field. In industry, the length of mass-produced components when they emerge from a production-line is often found to vary about its target value according to a normal distribution, and some of the standard I.Q. tests produce scores which follow a normal distribution. Every field of application of statistics gives rise to examples of data that follow the normal distribution to at least a good degree of approximation.

Examples also exist of variates which are not themselves normal, but can be made so by a suitable *transformation* of their scale of measurement. When an insecticide is applied experimentally to insects, the concentration x of insecticide needed to kill any particular insect—what is called the tolerance of that insect—is not constant for all insects in the experiment; in fact it often happens that when the frequency of kill is plotted against concentration, it gives a skew curve like Fig. 6.2(a). By plotting log x, instead of x, on the horizontal axis, the curve can be normalized (Fig. 6.2(b)), and we say that the original data on concentration, x, and kill were **log-normally** distributed, since log x is normal. (Note that $x = 1$ in Fig. 6.2(a) corresponds to log $x = 0$ in (b).) The log-normal distribution is also found in metallurgy, where specimens have been cast from an alloy and are subjected to a test for a time x before showing cracks or other symptoms of metal fatigue.

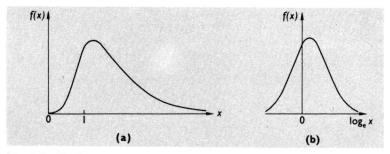

Fig. 6.2   The log-normal distribution: (a) shows $f(x)$ plotted against x and (b) shows $f(x)$ plotted against $\log_e x$. (The distribution in (b) is normal.)

The normal distribution can also be used to approximate to several other standard distributions, given suitable conditions on their parameters: we say that these distributions **converge to normality**. It is even possible to approximate to discrete distributions in this way, in spite of the normal being itself continuous. The *convergence* is in the sense that, when the necessary conditions are satisfied, a frequency graph for the actual distribution will be very similar indeed in shape to the normal which is used as its approximation. Hence statistical summary measures and significance tests (Chapters 7, 8, 9) worked out for the approximating normal distribution may be used, instead of carrying out the calculations for the actual distribution itself. This is often easier to do, and also has the advantage that separate methods and theory need not be developed for each distribution. Examples of the required conditions are that a *binomial* distribution can be approximated, for large

n, by $\mathcal{N}$ (np, npq), the normal with the same mean and variance as the binomial; and a *Poisson*, for large λ, is very close to $\mathcal{N}$ (λ, λ). For the binomial, the approximation is satisfactory for $n > 100$, and also for smaller values of n (say > 30) if p is near to $\frac{1}{2}$. (For $p = \frac{1}{2}$, the binomial is in any case a symmetrical distribution, but this does not automatically imply normality. The rules for approximation tell us how large n needs to be before a bar-diagram of the binomial would be covered very closely by the curve of its approximating normal.) For the Poisson distribution, the approximation is a good one when λ is 5 or greater.

It should be realized that when we summarize a set of data by quoting its mean and standard deviation, we are in a sense assuming it to be, more or less closely, normally distributed: for if we know that the data follow a binomial or a Poisson distribution, other parameters are appropriate (n and p, or λ respectively), and if we know the distribution to be rather skew, or of no particular shape at all, there is often a case for using medians and the measures of dispersion related to them. It *is* very common to summarize data by mean and standard deviation, but a moment's thought should be given to the appropriateness of this.

THE CENTRAL LIMIT THEOREM

This is the most common justification for using the normal distribution, and to appreciate it we need the idea of a ***sampling distribution***. Let us suppose that we take a sample of N observations, at random, from *any* frequency distribution, not necessarily normal: we ask of it only that the frequency $f(x)$ shall tail off sufficiently rapidly as x increases for it to possess a finite mean and variance (in practice we could assume this for biological data in general). This sample will have a mean, $\bar{x}$. Suppose we now take a second sample, also of N observations, from the same original (*parent*) distribution; this too will have a mean, whose numerical value will be slightly different from the first $\bar{x}$ because the sample, although from the same parent population, consists of different members. As we go on taking samples, so we go on getting different numerical values for their means, and this set of numerical values represents the *sampling distribution of the mean*. At first sight we might suppose that the shape or pattern of this sampling distribution would depend very much on the shape of the parent distribution; but this is not so. It is a very remarkable mathematical result that the sampling distribution of the mean of samples from *any* parent distribution approaches a normal distribution very closely as sample size N increases. Further, we can say what the mean and variance of this sampling distribution of means will be: if μ, σ^2 denote the parent mean and variance, then the sample mean based on N observa-

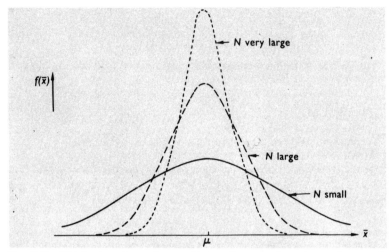

Fig. 6.3   Distribution of sample mean $\bar{x}$ about its true value μ for varying sample sizes N.

tions has the distribution $\mathcal{N}$ $(\mu, \sigma^2/N)$, this being an approximation satisfied more and more closely as N increases.

When the parent distribution is itself *normal*, the sampling distribution of means is *exactly* normal for all sample sizes N. This is the basis of the rule saying that the standard deviation of the mean of a sample is $\sigma/\sqrt{N}$, and shows that the precision of determination of a mean improves directly as the square root of the number of observations upon which it was based: its distribution becomes narrower as N increases, as shown in Fig. 6.3.

EXERCISES

(For Answers and Comments, see p. 165)

6.1   Find the unit (standard) normal deviate corresponding to:

(a) the value 5·00 in $\mathcal{N}$ (3.95, 2.25);
(b) the value 0·29 in $\mathcal{N}$ (0.50, 0.64);
(c) the value -0.47 in $\mathcal{N}$ (1.38, 1.21);
(d) the value -6.89 in $\mathcal{N}$ (-6.50, 0·04).

6.2   Draw rough graphs to show how the following pairs of normal distributions differ from each other:

(a) $\mathcal{N}(3, 1)$ and $\mathcal{N}(3, 4)$;
(b) $\mathcal{N}(0, 2)$ and $\mathcal{N}(6, 2)$;
(c) $\mathcal{N}(4, 1)$ and $\mathcal{N}(-4, 1)$;
(d) $\mathcal{N}(0, 1)$ and $\mathcal{N}(0.5, 0·1)$.

6.3 Using the data of Exercises 2.3 and 3.3, make a summary and draw a histogram to show how frequency is related to log x.

6.4 Which of the following binomial distributions could be approximated closely by a normal?
(a) the binomial $n = 8$, $p = \frac{1}{6}$;
(b) the binomial $n = 80$, $p = \frac{3}{5}$;
(c) the binomial $n = 50$, $p = \frac{1}{2}$;
(d) the binomial $n = 50$, $p = \frac{1}{10}$;
(e) the binomial $n = 300$, $p = \frac{9}{10}$;
(f) the binomial $n = 250$, $p = \frac{1}{1000}$.
State the means and variances of the appropriate normal distributions.

6.5 Sixty-four observations are selected at random from $\mathcal{N}(10, 25)$, and their mean $\bar{x}$ is calculated. What distribution will $\bar{x}$ follow?

6.6 Two hundred observations are selected at random from a distribution whose mean is 5 and variance 8. What distribution will the mean, $\bar{x}$, of these 200 observations follow? What could you say about $\bar{x}$ if only 20 observations were available instead of 200?

6.7 A radioactive source emits 2 particles per minute, on average. It is observed for N periods, each of 5 minutes, and in each period r, the number of particles emitted, is counted.
Can r be approximated by a normal distribution? If so, what are the mean and variance of this normal distribution?

6.8 Fertilizer is packed in (nominally) 5 kg bags. The packer has set his machinery to give bags of average weight 5·1 kg, and he knows that the distribution of bag weights will be normal with standard deviation 0·05 kg. A farmer buys 10 bags at a time. What is the distribution of mean weight of bags per purchase?

6.9 A measurement x cm is made in an experiment, and is repeated N times. The standard deviation of x is 0·5 cm. What should N be if the standard deviation of the mean measurement $\bar{x}$ is to be 0.1?

6.10 A set of weights, measured in lb, is distributed as $\mathcal{N}(10, 14·65)$. What will be the distribution of the same weights when they are expressed in kg? [1 lb = 0·4536 kg].

7

The Normal Distribution:
its use in Significance Tests

SIGNIFICANCE TESTING

Often a set of data is collected, or an experiment carried out, not simply with a view to summarizing the results and estimating suitable parameters but rather in order to *test an idea*. This idea or **hypothesis** may arise by purely theoretical reasoning, or it may have been suggested by the results of earlier experiments. In Chapter 5 we argued that if a set of data, such as the counts of numbers of insects or of radioactive particles, had been generated by a random process, the data should conform to a Poisson distribution. Thus if a set of such data does follow the Poisson distribution, it could have been generated in this way; but if it does not, then the hypothesis about a random process is not a reasonable one, because a deduction made from it, namely that the data ought to follow a Poisson distribution, is false.

Suppose, as a second example, that a series of experiments has been carried out at a research centre to determine the yield of strawberry plants when given a complete fertilizer. An experimental unit consisted of four plants of a standard variety; let us assume that the average yield per unit was 3 kg and that the variation in yield figures among the units followed a normal distribution. So in examining the results of a new experiment on the same variety, with units of four plants and the same fertilizer but carried out elsewhere, we may feel that the yield ought to be the same, and therefore set up the hypothesis that our resulting yield in this new experiment is normally distributed with mean 3. In practice (see Chapter 12) it would be more common to have *several* different fertilizer treatments in the experiment, each treatment applied to several different units, and we would postulate that the average yields under all the treatments were *the same*, without specifying any actual value. Thus in order to compare two of these treatments, we require to examine the difference between the means of two samples from normal populations, setting up the hypothesis that this difference is 0.

Obviously hypotheses cannot be set up on purely statistical grounds, but will embody any biological knowledge available at the outset of an experiment, or before data are collected. (One modern school of thought among statisticians assumes that at the outset a probability distribution can be given for the parameters in the hypothesis being tested, and that this probability distribution is modified in the light of the experimental results. Since the specification of these *prior probabilities* is often difficult, even arbitrary, in biological work, we shall not pursue these ideas in this book.) In any event, no experimenter or statistician would waste time testing statistically the reasonableness of a hypothesis which would be indefensible biologically if accepted. We shall see shortly that an element of uncertainty must be attached to acceptance or rejection; therefore we can never *prove* any particular hypothesis *correct*, but only show that it is (or is not) a *reasonable explanation* of the data available.

Any set of data gathered by observation or experiment inevitably constitutes a small sample from a large population, and so is subject to the random variation described in earlier chapters. Clearly, therefore, we cannot expect such a set of data to conform *exactly*, in the numerical sense, to any hypothesis, however well thought out. Hence we can examine only whether a set of data conforms to a hypothesis *sufficiently closely*, in a statistical sense to be defined; and if it does so, we accept the hypothesis as reasonable not only for that particular set of data but for the general population from which it was drawn (cf. Chapter 1). Even at this stage, assuming conformity to hypothesis, it is quite possible for someone else to put forward another hypothesis, perhaps a fundamentally different one, which may also be compatible with the data; when this happens, eventual choice between the two hypotheses must be made on biological grounds. After testing conformity, the statistician cannot make any further objective contribution to the argument. It should be noted, however, that if a new hypothesis is constructed *on the basis* of the available data, rather than before seeing them, it cannot command the same credence; there is no guarantee whatever that a repeat of the experiment will give data conforming to the new hypothesis, for this might have been fundamentally influenced by a particular accidental pattern present in the set on which it was based.

There is, of course, nothing to prevent such a new hypothesis being examined by a further experiment.

THE ROLE OF THE NORMAL DISTRIBUTION

The simplest test of significance will now be described in detail, not because others are less important but because it illustrates well the logical steps followed in any test. It is based on the normal distribution in standard form, $\mathcal{N}(0, 1)$, described in Chapter 6.

Test 7.1   A single observation is given, whose numerical value is represented by the symbol d. The hypothesis under test, called the ***Null Hypothesis (N.H.)***, is that d has been picked at random from $\mathcal{N}(0, 1)$.

On this hypothesis, the *possible* values of d are unrestricted, since a normal variate may take any value whatever, positive or negative, large or small. But if we set $\mu = 0$ and $\sigma^2 = 1$ in Fig. 6.1, it is clear at once that the frequencies of variate-values (that is, d-values) some distance away (positively or negatively) from the mean, 0, are small, and of values a considerable distance away, very small indeed. There exist various tables of the normal distribution; the most useful of these for present purposes gives, for each value of d, the cumulative frequency of values of the variate $\mathcal{N}(0, 1)$ that are less than or equal to d. Mathematically, this cumulative frequency is represented by the area under the frequency curve from its beginning up to d; and since the total area under the curve $\mathcal{N}(0, 1)$ is equal to 1, the cumulative frequency up to d will be the *proportion*, in the whole population, of members which have values less than or equal to d. This, as we have seen in Chapter 4, is equated to the *probability* of finding a member whose value is less than or equal to d when choosing *one* at random from the population.

Fig. 7.1(a) shows three points, $d = -1.96$, $d = -2.58$ and $d = -3.29$, which have the following properties: the probability (cumulative frequency) of values of d less than -1.96 is 0.025, or $2\frac{1}{2}\%$; of d less than -2.58 is 0.005, or $\frac{1}{2}\%$; of d less than -3.29 is 0.0005 ($\frac{1}{20}\%$). In other words, when choosing a single member at random from $\mathcal{N}(0, 1)$ we have a probability of $2\frac{1}{2}\%$ that it will be numerically less than -1.96, a probability of $\frac{1}{2}\%$ that it will be less than -2.58, and of $\frac{1}{20}\%$ that it will be less than -3.29. Since $\mathcal{N}(0, 1)$ is a curve symmetrical about 0, the probability of obtaining a value of d greater than $+1.96$ is also 0.025 ($2\frac{1}{2}\%$); so that the total probability of obtaining a value of d which is either less than -1.96 or greater than $+1.96$ is 0.05, or 5%. This implies that the probability of obtaining a value of d between -1.96 and $+1.96$ is 0.95, or 95%. Fig. 7.1(b) illustrates this.

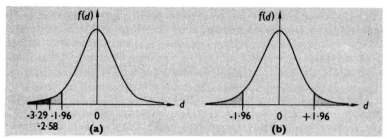

Fig. 7.1 (a) Values $d = -3.29$, -2.58, -1.96 in the $\mathcal{N}(0, 1)$ distribution, whose cumulative frequencies are respectively $\frac{1}{20}\%$, $\frac{1}{2}\%$, $2\frac{1}{2}\%$. (b) Values $d = \pm 1.96$ which enclose the central 95% of the $\mathcal{N}(0, 1)$ distribution.

Thus when a single member is chosen at random from $\mathcal{N}(0, 1)$, there is a probability 0·95 that its value d will lie between -1.96 and $+1.96$, and a probability 0·05 that d will be either less than -1.96 or greater than $+1.96$. The 95% central bulk of the $\mathcal{N}(0, 1)$ population extends from -1.96 to $+1.96$; so we expect, with probability 0·95, that one single value picked at random from $\mathcal{N}(0, 1)$ will lie in this range. Note the use of *probability* in this latter statement: on the hypothesis (the N.H.) that the population is $\mathcal{N}(0, 1)$ we make the assertion that a single member picked at random from this population *will have* its d value between -1.96 and $+1.96$; but this assertion is only true with 95% probability. If we go on picking such members, and make the same assertion every time, we shall in the long run be proved right 95% of times. But if the value of d for one particular member does *not* lie in this range, what can be said? Either we have picked an *unlikely* member, one of the sort that we expect with probability only 0·05; or our N.H. is incorrect, so that we are not in fact selecting from a $\mathcal{N}(0, 1)$ population and d does not therefore have to lie in the range -1.96 to $+1.96$.

When carrying out a significance test, we make the rule that if $|d| > 1.96$ (i.e. $d > +1.96$ *or* $d < -1.96$: $|d|$ stands for numerical value ignoring sign—the *modulus* of d), the N.H. that d comes from $\mathcal{N}(0, 1)$ is rejected, and the value of d is called *significant at the 5% level. Thus the essential step of logic in a significance test is to assume that when an unlikely value arises it is the N.H. which is at fault.* No value of d is completely impossible, so we never show conclusively that the N.H. is wrong, but only obtain more or less strong evidence to suggest that it might be. If the value of $|d|$ is > 1.96, the evidence is fairly strong; however when $|d| > 2.58$ the evidence is stronger, and when $|d| > 3.29$ it is stronger still, because if the N.H. is true, values of $|d| > 2.58$ should occur with probability only 0·01 (1%), and $|d| > 3.29$ with probability only 0·001 $\left(\frac{1}{10}\%\right)$.

SUMMARY OF TEST 7.1   *Null Hypothesis*: the given value of d has been picked at random from $\mathcal{N}(0, 1)$. Test: inspect the value of d, and if it is less than -1.96 or greater than $+1.96$ (i.e. if $|d| > 1.96$) reject the N.H. at the 5% level of significance. If $|d| > 2.58$, reject the N.H. at the 1% level, and if $|d| > 3.29$ reject the N.H. at the 0.1% level. (Other values of d are given in Table I.) The symbol z is often used for a standard normal variate, but we shall use d throughout this book.

Choice of levels of significance

The values $P = 0.05$, 0.01 and 0.001 used above are the commonly used levels of significance; we are not willing to reject a N.H. when the value of d being examined is more likely than once-in-twenty (i.e. probability 5%) to have arisen if the N.H. were true. It is also useful to have the stricter standards of unlikeliness, once-in-100 (1%) and once-in-1000 ($\frac{1}{10}$%) to be applied when it would be particularly undesirable to reject a N.H. that was, in fact, true. (It is often suggested that, when comparing two sets of data by the method to be described in Test 8.2, only those differences which appear significant at 5% are worth following up in subsequent experiments, those which are significant at 1% can be safely reported in publications, and those which are significant at 0.1% can be regarded as soundly established—always assuming that they make biological sense.)

Test 7.2   A single value x is given, from a normal population whose variance is known to be σ^2; test the hypothesis that the mean of this population is equal to some specified value μ. The test is to calculate $(x - \mu)/\sigma = d$, and examine the numerical value of this in the same way as for d in Test 7.1. We are employing here the property described in Chapter 6, that if x is $\mathcal{N}(\mu, \sigma^2)$ then $(x - \mu)/\sigma$ is $\mathcal{N}(0, 1)$. So if d is not an acceptable member of $\mathcal{N}(0, 1)$, we can equally well say that x is not an acceptable member of $\mathcal{N}(\mu, \sigma^2)$.

EXAMPLE 7.1

(*i*) Suppose that a Null Hypothesis has been set up which states that an observation x is distributed according to $\mathcal{N}(7.25, 1.69)$. A value $x = 3.35$ is then observed; is this observation consistent with the hypothesis?

Calculate

$$d = \frac{x - \mu}{\sigma} = \frac{3.35 - 7.25}{\sqrt{1.69}} = -\frac{3.90}{1.30} = -3.0$$

This lies between those d-values required for significance at 1% and 0.1%, which are respectively 2.58 and 3.29; it is thus less likely than once-in-100 to arise by chance if the N.H. is true (though not so unlikely as once-in-1000) and we may reject the N.H. at the 1% level of significance (though not at 0·1%). The observed value of x is described as *significant at 1%*.

(*ii*) A Null Hypothesis states that x should be $\mathcal{N}(4\cdot50, 0\cdot36)$; an observed value of x is 5·57. Thus

$$d = \frac{5\cdot57 - 4\cdot50}{\sqrt{0\cdot36}} = \frac{1\cdot07}{0\cdot60} = 1\cdot78$$

which is less than the value of d required for significance at the 5% level (namely 1·96) and so in this case we shall accept the N.H.

(*iii*) A Null Hypothesis states that x should be $\mathcal{N}(0\cdot40, 0\cdot09)$, and a value $x = -0\cdot80$ is observed. This time,

$$d = \frac{-0\cdot80 - 0\cdot40}{\sqrt{0\cdot09}} = -\frac{1\cdot20}{0\cdot30} = -4\cdot0$$

greater than $-3\cdot29$ in numerical value so that the N.H. is rejected at the 0·1% level of significance.

(*iv*) A N.H. states that x is $\mathcal{N}(15\cdot09, 1\cdot44)$, and an observation $x = 17\cdot43$ is made. Therefore

$$d = \frac{17\cdot43 - 15\cdot09}{\sqrt{1\cdot44}} = \frac{2\cdot34}{1\cdot20} = 1\cdot95$$

a value which is just on the borderline of significance at the 5% level. Now this does not provide strong evidence for rejecting the N.H., but neither does it support its acceptance with any great confidence. In such a case, the statistical test has not made any very definite contribution towards our attitude to the N.H. (except perhaps to prevent us from accepting it unreservedly) and before it can do so we must ask for more data: the test then will be based on a value of $\bar{x}$, as in Test 7.3 below, and not just on a single value of x.

EXAMPLE 7.2

A cytologist has studied chromosome sizes in a large number of healthy persons, and found that for one particular chromosome the ratio of its long arm to its short arm is normally distributed with mean value 1·75 and variance 0·0025. He measures the same ratio (long arm/short arm) for the same chromosome in a patient suspected to have some genetic abnormalities; the value of the ratio measured is 1·61. Shall he classify the patient as healthy or not?

Test 7.2 applies here, for we are given a value, $x = 1\cdot61$; and the N.H.

states that in a healthy patient x should be taken from the distribution $\mathcal{N}(1\cdot75, 0\cdot0025)$. Thus

$$d = \frac{1\cdot61 - 1\cdot75}{0\cdot05} = -\frac{0\cdot14}{0\cdot05} = -2\cdot80$$

This is greater in numerical value than $-2\cdot58$, and so is significant at 1%; at the 1% level, therefore, we may reject the N.H. that x was taken from $\mathcal{N}(1\cdot75, 0\cdot0025)$, and we are at liberty to say that this patient cannot reasonably be classed as healthy. Of course, it is quite possible that in spite of the statistical evidence, the cytologist will not want to classify a patient on the basis of one single observation, and he would take more chromosome samples and calculate more arm ratios; the mean value of these ratios would then be tested by Test 7.3. Alternatively he would measure other possible indicator characteristics as well as the arm ratio.

Test 7.3   A sample of N observations, $x_1, x_2, x_3, \ldots, x_N$ is given from a normal population whose variance is known to be σ^2.

Test the Null Hypothesis that the mean of this distribution is μ. The test applies a property noted in Chapter 6 (p. 42), namely that the *mean* of a random sample of N observations from $\mathcal{N}(\mu, \sigma^2)$ will have the distribution $\mathcal{N}(\mu, \sigma^2/N)$. Therefore the N.H. can be re-phrased '$\bar{x}$ has been picked at random from $\mathcal{N}(\mu, \sigma^2/N)$'. The method of Test 7.2 may now be applied to $\bar{x}$, by calculating $d = (\bar{x} - \mu)/\sqrt{\sigma^2/N}$ and then examining whether d is $\mathcal{N}(0, 1)$.

EXAMPLE 7.3

A sample of eight observations, drawn at random from a normal distribution whose variance is known to be 9, has a mean of $5\cdot75$. Test the N.H. that the mean of the distribution from which the sample was drawn had the value 4.

The mean of eight observations from $\mathcal{N}(4, 9)$ will itself have distribution $\mathcal{N}(4, \frac{9}{8})$, i.e. $\mathcal{N}(4, 1\cdot125)$. So if the N.H. is true, the sample mean $5\cdot75$ will have this distribution $\mathcal{N}(4, 1\cdot125)$. Thus

$$d = \frac{5\cdot75 - 4\cdot00}{\sqrt{1\cdot125}} = \frac{1\cdot75}{1\cdot06} = 1\cdot65$$

which is not significant at the 5% level (being less than $1\cdot96$), so that we can accept the sample mean as a member of $\mathcal{N}(4, 1\cdot125)$ and therefore we can accept the N.H. that the mean of the distribution from which the sample was drawn was 4.

EXAMPLE 7.4

After incubation for 24 hours at 18 °C, spores of a particular species of fungus are examined under a microscope. Over a long period of study the average length of their germ-tubes has proved to be 8·2 scale-divisions, and the variance of length 0·052. A new incubator is installed, and spores of the same species incubated in it for 24 hours at 18 °C. A random sample of 20 of these spores is selected, and examined in the customary way under the microscope. The mean of the 20 germ-tube lengths is 8·32 scale-divisions. Test the N.H. that the growth rate over the 24 hours is unchanged.

Test 7.3 may be applied, on the N.H. that the growth rate is still $\mathcal{N}(8\cdot2, 0\cdot052)$. The mean of 20 observations from $\mathcal{N}(8\cdot2, 0\cdot052)$ will then follow the distribution $\mathcal{N}(8\cdot2, 0\cdot052/20)$, and we therefore test whether $\bar{x} = 8\cdot32$ is an acceptable member of this latter distribution.

$$\frac{\bar{x} - \mu}{\sqrt{\sigma^2/N}} = \frac{8\cdot32 - 8\cdot20}{\sqrt{0\cdot052/20}} = \frac{0\cdot12}{\sqrt{0\cdot0026}} = \frac{0\cdot12}{0\cdot051} = 2\cdot35$$

which is significant at the 5% level. We therefore reject the N.H. at the 5% significance level, and consider that we have evidence against the theory that growth rate is unchanged.

APPROXIMATE NORMALITY

As explained in Chapter 6 (p. 40), many variates are *approximately normal* in the sense that under suitable conditions their distributions will be very close indeed to normality in shape even though not exactly normal. If we use the above Tests 7.1, 7.2, 7.3 for these variates, we obtain very good indications of whether or not to accept such Null Hypotheses as are set up; but in so far as the distributions are not exactly normal, the significance levels applied will not be exactly 5%, 1% or 0·1% but only approximately so. Therefore we are bound to refer to this class of test as *approximate*, in contrast to the *exact* tests previously described, although unless we are very careless about the conditions under which the tests are applied the approximation will be a very close one.

We noted at the end of Chapter 6 (p. 41) that much of the importance of the normal distribution derives from the Central Limit theorem. Under the conditions of that theorem we may apply Test 7.3 to the *means* of samples drawn at random from *any* distribution, provided these samples are *large*.

For example, if $\bar{x}$ is the mean of $N = 500$ observations from any distribution, whose mean $\mu = 10$ and variance $\sigma^2 = 50$ are given, the sample is such a large one that the distribution of $\bar{x}$ will be very close indeed to $\mathcal{N}(10, \frac{50}{500})$, i.e. $\mathcal{N}(10, 0\cdot1)$, and this can be used to test the significance of a given numerical value of $\bar{x}$. It is wise to require $N > 100$ before applying this approximation, though if the distribution from which the sample is drawn is itself not too far from normality, the method will work well enough for smaller samples.

The mean of N observations from a Poisson distribution may also be tested in this way; in this case we know that the distribution has mean and variance both equal to λ, so that the mean $\bar{r}$ will be approximately $\mathcal{N}(\lambda, \lambda/N)$. The size of N needs to be large if λ is small; but when λ is as large as 5 the Poisson distribution is itself not far from normality so that the size of N need not be greater than, say, 30.

Means of observations from other discrete distributions also admit of similar treatment, and as for the Poisson case the sample size N necessary for satisfactory approximation will depend on the shape of the original distribution.

The Poisson distribution with mean > 5: a single observation

EXAMPLE 7.5

The average rate of emission of radioactive particles from a source was measured over a long period, and found to be 10 particles per unit time. After an experimental treatment had been applied to the source, a further sample was examined and emitted 17 particles in unit time. Test the N.H. that the rate of emission is unchanged.

The variate r, the number of emissions per unit time, follows the Poisson distribution, but since its average is greater than 5 we may obtain a close approximation to its distribution by employing the normal with the same mean and variance, i.e. $\mathcal{N}(10, 10)$. Thus the new observation, 17, has to be tested as a single member of $\mathcal{N}(10, 10)$ in the manner shown in Examples 7.1, 7.2.

We have

$$d = \frac{17 - 10}{\sqrt{10}} = \frac{7\cdot00}{3\cdot16} = 2\cdot21$$

which is certainly significant at the 5% level, and so the N.H. is rejected: it is *not* reasonable to say that the emission rate is unaltered.

In exactly the same way, Test 7.3 may be applied to testing hypotheses about the means of Poisson distributions.

The Binomial distribution with large samples

EXAMPLE 7.6

A plant breeder knows that if the skin colour of a tomato produced by crossing two given parents is determined by a single factor, three-quarters of a batch of seedlings will be red and the remainder striped. Upon growing 100 such seedlings, he observes that 64 of them are red and 36 striped. Can the hypothesis of a single factor be accepted?

This requires a test of whether the observations on the sample of $N = 100$ seedlings are consistent with the hypothesis that $p = \frac{3}{4}$ in the population. The distribution of r, the actual number of red seedlings observed in the sample of N ($r = 64$ in the example), is binomial; and as we saw in Chapter 4 (p. 28) the variance of r is therefore equal to Npq. But we are interested in the *proportion*, $\hat{p} = r/N$, rather than in the number r itself: in problems involving the binomial distribution it is often more convenient to consider proportions in preference to numbers. (Note that since the estimate of p is based on *all* N observations made, the calculation proceeds as though we had *one* binomial sample, of parameters N, p.)

If all the measurements made on a variate are divided by a constant number, the standard deviation is divided by the same constant and, in consequence, the variance by the square of the constant. So when we study $\hat{p}$, we note first that it is equal to the variate r divided by the constant number N: hence the variance of $\hat{p}$ is the variance of r divided by N^2. This is $Npq/N^2 = pq/N$. Further, if N is not too small and p not too near 1 or 0, both of which conditions are adequately satisfied here, we may assume both r and $\hat{p}$ to be approximately normally distributed.

This gives the useful approximation that, under the conditions just stated, *a proportion $\hat{p}$ is approximately $\mathcal{N}(p, pq/N)$*.

Note that in Example 7.6, we are told that $p = \frac{3}{4}$, and so may use this value of $\hat{p}$ in the statement of the Null Hypothesis, which now reads 'the observed value $\hat{p}$ is a member of $\mathcal{N}(\frac{3}{4}, \frac{3}{4} \times \frac{1}{4} \div 100)$, i.e. $\mathcal{N}(\frac{3}{4}, \frac{3}{1600})$'.

The observed $\hat{p} = r/N = 64/100 = 0.64$. Therefore we test as $\mathcal{N}(0, 1)$

$$ d = \frac{\hat{p} - \frac{3}{4}}{\sqrt{\frac{3}{1600}}} = \frac{0.64 - 0.75}{0.043} = \frac{-0.11}{0.043} = -2.56 $$

significant at 5% (almost at 1%), which leads us to reject the N.H. that $p = \frac{3}{4}$, and so to look for a less simple genetic law to explain the results.

ONE- AND TWO-TAILED TESTS

In all the examples quoted so far we have used the idea suggested by

Fig. 7.1(b), that *both* extremities of the normal curve (*d* greater than +1·96 or less than −1·96) contain *unlikely* values of *d*, whose appearance in the result of a significance test makes us doubt the validity of the N.H. But if we do reject the N.H., what hypothesis shall we accept? Until now, we have tacitly accepted as *Alternative Hypothesis* (or A.H.) in these cases that if our observed values or means do not come from a normal distribution with the specified value of μ, then they come from a distribution with some other value of μ which is not specified at all, and is equally likely to be higher or lower than that stated in the N.H. We shall refer to this A.H. as the *vague* one; and in all these cases we have rejected extreme values in *either* direction as in Fig. 7.1(b), so that the tests so far described may be called *two-tailed* ones because both ends (or *tails*) of the normal curve are rejected.

But sometimes the A.H. is not just the vague one: it actually specifies something about the alternative value of μ which is to be accepted if the value of μ in the N.H. is rejected. The A.H. may set a numerical value to μ, or it may say that the alternative value of μ is greater than that in the N.H. without actually giving a numerical value (or, of course, that the alternative value is less than that in the N.H.). The first of these possibilities is included in the other two (and the third dealt with in a way very obviously similar to the second) so that we can illustrate the general method for one-tailed tests by a single example.

EXAMPLE 7.7

A pathologist knows that mycelial colonies (*spots*) of a common species of fungus, when incubated under standard laboratory conditions for a fixed period of time, grow to an average diameter of 26 units, and that the distribution of diameters in a large population of mycelial spots is $\mathcal{N}(26, 16)$. A closely related species, visually similar, has colony diameter distributed as $\mathcal{N}(30, 16)$. In a given sample of 50 spots, the mean diameter proves to be 29·4 units; the null hypothesis is that the sample is from $\mathcal{N}(26, 16)$, but if not then it must be from $\mathcal{N}(30, 16)$.

Fig. 7.2 shows the two possible normal distributions for this example,

Fig. 7.2 Distributions of colony diameters, $\mathcal{N}(26, 16)$ and $\mathcal{N}(30, 16)$ (Example 7.7).

the one to the left being the N.H. upon which calculations are based: on this, the mean of 50 observations will be $\mathcal{N}(26, \frac{16}{50})$, and so we test 29·4 as a possible member of $\mathcal{N}(26, \frac{16}{50})$. This gives

$$d = \frac{29·4 - 26·0}{\sqrt{0·32}} = +\frac{3·40}{0·57} = +6·0$$

which is extremely large, and so is significant at 0·1%. Therefore we reject the N.H. and automatically accept the A.H. This is reasonable, here, because the A.H. specifies a larger mean than the N.H. and this sample did have a larger mean. But what if it had had a mean considerably smaller than 26? The A.H. under those circumstances would have been more unreasonable than the N.H., and could not sensibly have been accepted: in fact, *all* negative values of d are more likely to occur on the N.H. than on the A.H., so we cannot in this situation reject the left-hand tail. All the region of rejection of values of the mean calculated on the N.H. must be at the *right-hand* end, as shown in Fig. 7.3; we can only reject those values that can be better explained by the A.H. So

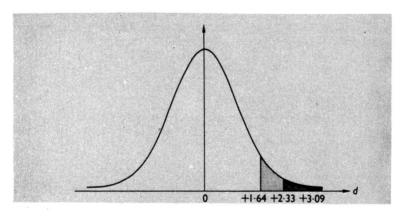

Fig. 7.3   Values of d for one-tailed significance tests at 5%, 1%, 0·1% levels.

if we wish to make a test at the 5% significance level, we must use that value of d which will exclude 5% of the area under the curve at the *upper end alone*, namely $d = +1·64$. (This is of course the numerical value of d, taken both positively and negatively, which in a two-tailed test would exclude 10% of the total area; and so it appears as a 10% entry in most standard tables.) The corresponding values of d for one-tailed tests at the 1% and 0·1% levels are 2·33 and 3·09 respectively.

EXERCISES

(For Answers and Comments, see p. 165)

7.1   For each of the parts (a)–(d) of Exercise 6.1 (p. 42) test the hypothesis that the given value came from the normal distribution quoted [i.e. that 5·00 came from $\mathcal{N}(3\cdot95, 2\cdot25)$ etc.].

7.2   Sixty-four observations are selected at random from a normal distribution whose variance is 25. Their mean is calculated, and found to be 11.1. Test the hypothesis that the true value of the population mean is 10.

7.3   Twenty guinea-pigs from specially bred laboratory stock are fed on a standard diet for a fortnight from birth. Their mean increase in weight is 28 units. In the past, the mean for a very large number of similar animals has been 29·8 units and the variance of weight-increase 25 units. It is thought that these 20 animals may be different because they have been housed in a newly-installed pen. Test whether their weight increase does differ from the result obtained for the large population.

7.4   Two hundred observations are selected at random from a distribution whose mean is thought to be 5 and variance known to be 8. The mean of the 200 observations is 4·77; test whether the hypothesis, that the population mean is 5, is acceptable.

Repeat the above for a sample size 20 instead of 200.

7.5   The numbers of isopods in each of 50 sampling quadrats were recorded, and found to follow a Poisson distribution; the mean number of isopods per quadrat was 2·2. Test the hypothesis that the true mean, over the whole area from which the samples were drawn, is 2·0.

7.6   A Poisson variate is thought to have a mean of 6·25. If a single observation took the value 11, would it be reasonable to assume that it had come from this distribution?

7.7   In the problem of Exercise 4.4 (p. 30) test the hypothesis that the proportion of affected chromosomes is $\frac{1}{2}$.

7.8   A group of 10 strawberry plants is grown in ground treated with a chemical soil conditioner, and the mean yield per plant is 114 g. Experience has shown that when the same variety of strawberry is grown under similar conditions, but with no soil conditioner, the mean has been 110 g and the variance 84. Test whether it can reasonably be claimed that the soil conditioner had a beneficial effect on yield.

7.9   A standard drug used to relieve pain after a certain type of operation has been successful with 80% of patients. A new drug is tried, in the belief that it will be more successful, on 200 patients, and 170 of these gain relief. Is the belief in the new drug justified?

7.10   In Exercise 7.9, what could be said if (a) 152 or (b) 140 of the patients had gained relief?

7.11 The weights of a large number of children in the same age-group in a region have been measured, and the mean and variance in this population are respectively 102 lb and 49 lb². If 100 children from an adjacent region are weighed, and their mean weight is 99 lb, is there any evidence that the population in the second region differs from that in the first? If only 5 children had been available, how would you have examined their weight records?

8

Other Tests of Significance

STUDENT'S t

In Test 7.3 we were given a random sample of N observations, whose mean was $\bar{x}$. We assumed that this sample came from a normal distribution, and that we *knew* the variance σ^2 of the distribution. The Null Hypothesis to be tested was that the distribution had mean μ. Often, σ^2 is not known, but instead we must use an *estimate* s^2 of the variance, calculated from the sample by the method of Chapter 3. In the expression $d = (\bar{x} - \mu)/\sqrt{\sigma^2/N}$ of Test 7.3, there was only one item, $\bar{x}$, which varied from sample to sample. By analogy we now consider $t = (\bar{x} - \mu)/\sqrt{s^2/N}$ which contains *two* variable quantities, $\bar{x}$ and s^2: its distribution cannot therefore be the same as that of d. W. S. Gosset (who wrote under the pen-name *Student*) studied the distribution of t and found that it had a symmetrical frequency curve, bell-shaped like the normal, with mean o and a variance which depended on the size N of the observed sample. In fact the parameter of the t-distribution is not exactly N, but is the divisor $(N-1)$ in the expression for s^2: this value $(N-1)$ is called the **degrees of freedom** (d.f.) of s^2 and of t, and the expression $(\bar{x} - \mu)/\sqrt{s^2/N}$ is said to be *distributed as t with $(N-1)$ degrees of freedom*, often abbreviated to $t_{(N-1)}$. Those values of $t_{(N-1)}$ which exclude the extreme 5% or 1% or 0.1% of the area under its frequency curve can be found, to be applied in significance testing in exactly the same way as we have already used d; but because the exact shape of the t-distribution depends on its d.f., a t-table (Table I) needs to contain a row of these values for each of the d.f. $1, 2, 3, 4, \ldots$. Since t is one of the many distributions which converge to normality, in the manner discussed in Chapter 6, most tables quote values of *t* up to 30 d.f. only, *t* being very close to $\mathcal{N}(0, 1)$ for degrees of freedom above 30. The shapes of the curves of $t_{(6)}$, $t_{(20)}$ and $\mathcal{N}(0, 1)$ are compared in Fig. 8.1.

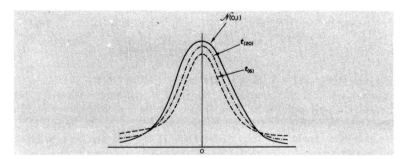

Fig. 8.1   Comparative shapes of the distributions $t_{(6)}$, $t_{(20)}$, $\mathcal{N}(0, 1)$.

Test 8.1   A random sample of N observations is given, from a normal distribution of unknown variance, and the sample mean $\bar{x}$ and variance s^2 are calculated. To examine the N.H. that the true mean of the normal distribution is μ, test $t_{(N-1)} = (\bar{x} - \mu)/\sqrt{s^2/N}$ as *t* with $(N-1)$ degrees of freedom.

EXAMPLE 8.1

Test whether the following random sample of 10 observations could reasonably have been taken from a normal distribution whose mean was 0·6: [1·2, 2·4, 1·3, 0·0, 1·0, 1·8, 0·8, 4·6, 1·4, 1·3].

The sum of all ten observations is 15·8, and so $\bar{x} = 15\cdot8/10 = 1\cdot58$. Now $\sum_{i=1}^{10}(x_i - \bar{x})^2$, where x_i denotes a typical one of the ten observations, is calculated by using the equivalent formula (p. 20) $\sum_{i=1}^{10} x_i^2 - (\sum_{i=1}^{10} x_i)^2/10$. The sum of the squares of the x_i is

$$(1\cdot2)^2 + (2\cdot4)^2 + \cdots + (1\cdot3)^2 = 38\cdot58$$

and

$$\frac{(\sum_{i=1}^{10} x_i)^2}{10} = \frac{(15\cdot8)^2}{10} = \frac{249\cdot64}{10} = 24\cdot964$$

So

$$\sum_{i=1}^{10}(x_i - \bar{x})^2 = 38\cdot58 - 24\cdot964 = 13\cdot616$$

The sample variance s^2 is

$$\frac{1}{9}\sum_{i=1}^{10}(x_i - \bar{x})^2 = \frac{13\cdot616}{9} = 1\cdot513$$

Therefore test

$$\frac{\bar{x}-\mu}{\sqrt{s^2/N}} = \frac{1 \cdot 58 - 0 \cdot 60}{\sqrt{1 \cdot 513/10}}$$

as $t_{(N-1)}$, i.e. $t_{(9)}$, which gives

$$t_{(9)} = \frac{0 \cdot 98}{\sqrt{0 \cdot 1513}} = \frac{0 \cdot 98}{0 \cdot 389} = 2 \cdot 52$$

On consulting Table I, we see that the 5% point for $t_{(9)}$ is 2·26, or in other words that the central 95% of the t distribution with 9 d.f. lies between $-2 \cdot 26$ and $+2 \cdot 26$. The observed value 2·52 is greater than the figure in the Table, i.e. is outside the central 95% of the distribution; we therefore reject the Null Hypothesis at the 5% significance level: we *cannot* reasonably suppose that the sample was drawn from a distribution whose true mean was 0·6.

In the above example, no definite Alternative Hypothesis was specified, but had this been done we should have had to consider whether to carry out a two-tailed test as above, or a one-tailed test. If a one-tailed test were decided on, it would be performed in exactly the same manner as the illustration of Example 7.7; the 10%, 2% and 0·2% entries in Table I are employed, in the same way as described for d.

Difference between two means: variance not known

We saw in Chapter 6 that if a sample of N observations was chosen at random from a normal distribution whose variance was σ^2, then the sample mean $\bar{x}$ had variance σ^2/N; when a second sample, this time of M observations, is drawn from a normal distribution with the same variance, its mean $\bar{y}$, say, will have variance σ^2/M. Now the difference $(\bar{x}-\bar{y})$ contains two varying elements, namely $\bar{x}$ and $\bar{y}$, both of which contribute to its total variance: in fact the variance of $(\bar{x}-\bar{y})$ is $\sigma^2/N+\sigma^2/M$ (note the $+$ sign). If we set up the N.H. that $\bar{x}$ came from $\mathcal{N}(\mu_1, \sigma^2)$ and $\bar{y}$ from $\mathcal{N}(\mu_2, \sigma^2)$, we must test whether the observed difference $(\bar{x}-\bar{y})$ could reasonably be equal to $(\mu_1-\mu_2)$. When the results of a designed experiment are being analysed, as in Chapter 12, much the most common need is to see whether $\mu_1=\mu_2$, and thus to test whether $(\bar{x}-\bar{y})$ could be 0; the value of σ^2 is very unlikely to be known, but has to be estimated from the information provided by all the treatments in the experiment.

Test 8.2   Two samples are given, both drawn from normal populations having the same variance (which is unknown); their sizes are N, M respectively and their means $\bar{x}$, $\bar{y}$. Test the N.H. that both samples were drawn from the same population, i.e. had the same mean as well as the same variance.

First calculate a pooled variance, that is a variance based on both samples,

$$s^2 = \frac{1}{N+M-2}\left\{ \sum_{i=1}^{N} (x_i - \bar{x})^2 + \sum_{j=1}^{M} (y_j - \bar{y})^2 \right\}$$

Now test

$$\frac{\bar{x}-\bar{y}}{\sqrt{s^2(1/N+1/M)}} \quad \text{as } t_{(N+M-2)}$$

Notice that when we estimated a variance from a single sample, we took the sum of squares of deviations $\sum(x_i - \bar{x})^2$ and divided it by the degrees of freedom $(N-1)$. When we want to make an estimate based on two samples, we *add* the sums of squares of deviations for *both* samples, $\sum(x_i - \bar{x})^2 + \sum(y_j - \bar{y})^2$, and divide by the *sum* of d.f., $(N-1)+(M-1)$. This gives an estimate s^2 whose d.f. are $(N-1)+(M-1)$, i.e. $N + M - 2$. (We could extend this method to use three or more samples in the same way.)

When we have s^2, the variance of $(\bar{x}-\bar{y})$ is estimated as $s^2/N+s^2/M$.

EXAMPLE 8.2

Seven plants of wheat grown in pots and given a standard fertilizer treatment yield respectively 8·4, 4·5, 3·8, 6·1, 4·7, 11·2 and 9·6 g dry weight of seed. A further eight plants from the same source are grown in similar conditions but with a different fertilizer and yield respectively 11·6, 7·5, 10·4, 8·4, 13·0, 7·0, 9·6, 13·2 g. Test whether the two fertilizer treatments have different effects on seed production.

For the first sample $\bar{x}=6\cdot900$ and $\sum_{i=1}^{7} (x_i - \bar{x})^2 = 48\cdot88$; and for the second, $\bar{y}=10\cdot0875$ and $\sum_{j=1}^{8} (y_j - \bar{y})^2 = 39\cdot8685$. Note that we do not estimate the variances of x and y separately, but use only these sums of squares in the pooled estimate

$$s^2 = \frac{48\cdot8800+39\cdot8685}{13} = 6\cdot8268$$

The value of $(N+M-2)=7+8-2=13$ is the d.f. for t, the samples are assumed to be normally distributed (we can reasonably assume this for measurements such as crop yield), and the hypothetical value for $(\bar{x}-\bar{y})$ is 0.

So we have the test

$$t_{(13)} = \frac{\bar{x}-\bar{y}-0}{\sqrt{s^2(\frac{1}{7}+\frac{1}{8})}} = \frac{6\cdot900-10\cdot0875}{\sqrt{(\frac{15}{56})\times 6\cdot8268}} = \frac{-3\cdot188}{\sqrt{1\cdot8286}} = \frac{-3\cdot188}{1\cdot35} = -2\cdot36$$

On consulting Table I we find that this value is significant at 5%, so casting considerable doubt on the N.H. that the two original distributions, from which the two samples were taken, had the same mean. This in turn leads us to reject, at the 5% level, the hypothesis that the two fertilizer treatments had the same effect.

The t distribution must be used whenever σ^2 is not known and the estimate s^2 has to be employed instead. It still requires samples to be drawn from normal populations, but it does not require them to be large: our examples, with samples of only 10, 7 and 8 observations are quite commonplace.

However, when samples *are* large, so that t has more than 30 d.f., the $\mathcal{N}(0, 1)$ curve approximates so closely to t that we may use it instead. But it is wrong to use $\mathcal{N}(0, 1)$, that is to replace t by d, for tests on small samples when σ^2 is not known.

Difference between two means: variances known

If two samples of observations are taken from normal distributions whose variances are known, tests of differences between sample means become simpler. Suppose that a sample of N observations from a normal distribution with *known* variance σ_1^2 is available; its mean is $\bar{x}$. A further sample, of M observations, from another normal distribution with known variance σ_2^2 has mean $\bar{y}$. Let us set up the hypothesis that the first population has true mean μ_1, and the second has mean μ_2. On this hypothesis, the difference $(\bar{x} - \bar{y})$ follows the distribution $\mathcal{N}(\mu_1 - \mu_2, \frac{\sigma_1^2}{N} + \frac{\sigma_2^2}{M})$ exactly, whatever the sizes of N, M. A test of the hypothesis is thus based on standardizing the distribution of $(\bar{x} - \bar{y})$, and calculating

$$d = \frac{(\bar{x} - \bar{y}) - (\mu_1 - \mu_2)}{\sqrt{\sigma_1^2/N + \sigma_2^2/M}},$$

which is tested as $\mathcal{N}(0, 1)$. Since this test is very rarely used in its exact form, we will not give an example here; but note the similarity with Test 8.3 for large samples.

Unequal variances which are unknown

If in Example 8.2 we had not been able to assume that the unknown value of σ^2 was the same in both samples, the t test could not immediately have been applied. In fact the correct mathematical treatment in such situations has been a subject for much discussion, but some approximate forms of test can be stated which hold under limited conditions.

Test 8.2a   If M, N are neither very small (say > 20) nor very different, Test 8.2 can be used as an *approximate* one, with the same expression for pooled s^2, even though it is not known whether the two
populations have the same variance.

Test 8.3   The same conditions are given as in Test 8.2, save that
the two normal populations from which samples are drawn do *not* have
the same variance; and the N.H. is that the means of the two populations are equal. Calculate the mean $\bar{x}$ and variance s_1^2 of the first sample
based on N_1 observations, and the mean $\bar{y}$ and variance s_2^2 of the second
sample based on N_2 observations. Then

$$t_{(N_1 + N_2 - 2)} = \frac{\bar{x} - \bar{y}}{\sqrt{s_1^2/N_1 + s_2^2/N_2}}$$

is *approximately* distributed as t with $(N_1 + N_2 - 2)$ d.f. This test serves
for smaller sample sizes than Test 8.2a, though when N_1 and N_2 are
both small and nowhere near equal (e.g., $N_1 = 4$ and $N_2 = 9$) *no* simple
satisfactory test can be found.

EXAMPLE 8.3

Two samples from normal distributions gave, respectively, means of
4·29, 4·18, and variances of 6·89, 2·55, based on 10, 12 observations.
Can these samples be from distributions with the same mean?
    Test

$$\frac{4\cdot29 - 4\cdot18}{\sqrt{6\cdot89/10 + 2\cdot55/12}} \quad \text{as } t_{(20)}$$

i.e.

$$t_{(20)} = \frac{0\cdot11}{\sqrt{0\cdot6890 + 0\cdot2125}} = \frac{0\cdot11}{\sqrt{0\cdot9015}} = \frac{0\cdot11}{0\cdot95} = 0\cdot12$$

which is certainly not significant, so that we can accept equality of means.

Relaxation of the Normality condition for large samples

When N_1, N_2 are large, we may replace $t_{(N_1 + N_2 - 2)}$ by d in Test 8.3
without making the approximation worse, since $N_1 + N_2 - 2$ will be far
above 30 and t thus very close to $\mathcal{N}(0, 1)$. Also, even if the populations
of observations x, y are not normal, the means $\bar{x}$, $\bar{y}$ will be very nearly
normal by the Central Limit theorem, and their variances will be s_1^2/N_1
and s_2^2/N_2. So, approximately, for *large* samples and provided only that

they satisfy the Central Limit theorem, *any* two parent distributions may have the equality of their means tested by Test 8.3 with d instead of t.

TESTS OF VARIANCES

So far we have been concerned only with the *location* of single observations or of means; but the *scatter* or *variability* of a set of observations, measured by its variance, is a characteristic which often gives useful information (and is perhaps not studied as often as it should be). There are two useful tests: one (χ^2) examines whether a variance in a normal distribution has a specified value, the other one (F) tests equality of two variances.

The χ^2 (or *chi-squared*) distribution

This distribution has several uses in significance testing. Its original mathematical form is concerned with the squares of normally-distributed observations. If we have a number, N, of $\mathcal{N}(0, 1)$ observations, $\{x_i\}$, then the sum of their squares, $\sum_{i=1}^{N} x_i^2$, has the χ^2 distribution. More generally, if we have a sample of N observations, $\{x_i\}$, from a normal distribution of unknown mean and known variance σ^2, then the sum of squares $\sum_{i=1}^{N} (x_i - \bar{x})^2$, divided by the known σ^2, has the χ^2 distribution with $(N-1)$ degrees of freedom. The estimated variance of the sample, s^2, is (Chapter 3) $\sum_{i=1}^{N} (x_i - \bar{x})^2/(N-1)$; so the sum of squares is $(N-1)s^2$, and $[(N-1)s^2]/\sigma^2$ is therefore distributed as $\chi^2_{(N-1)}$. The χ^2 distribution, like t, depends on the sample size N in the sense that its degrees of freedom are $(N-1)$, so that a χ^2-table (Table II) must contain a row for each value 1, 2, 3, ..., of the d.f. However, it is not a symmetrical distribution; also, since $\sum(x_i - \bar{x})^2$ can never take negative values (squares must be positive), χ^2 is defined only over the range of values 0 to infinity, not to $-\infty$. A typical shape for the χ^2 curve is shown in Fig. 8.2; the

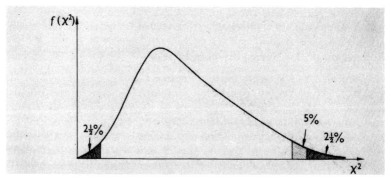

Fig. 8.2   The χ^2-distribution, showing one- and two-tailed 5% significance points.

actual position of the peak and the rates of increase and decrease of frequency on either side of it are all determined by the d.f. The hump moves steadily to the right as the d.f. increase, and the shapes of $\chi^2_{(3)}$ and $\chi^2_{(10)}$ are compared in Fig. 8.3.

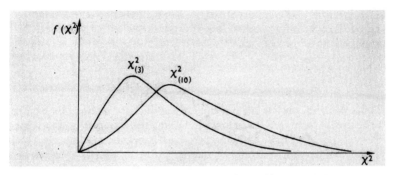

Fig. 8.3   Comparison of χ^2-distributions with 3 and 10 degrees of freedom.

There are two possible ways in which we may choose to exclude an extreme 5% part from the χ^2 distribution, illustrated in Fig. 8.2 by different degrees of shading. First, as for the normal and *t* distributions when used in two-tailed tests, we may choose to omit both the extreme $2\frac{1}{2}\%$ at the lower end and the extreme $2\frac{1}{2}\%$ at the upper end. Secondly, we may choose to omit only the upper 5%, accepting all values less than this, right down to o. This second, *one-tailed* procedure is needed more often than the first, and so the standard tables of χ^2 quote, for each d.f., the values of χ^2 *above* which lie only 5%, or 1%, or 0·1% of the total area. So when we want a two-tailed procedure, we need the $2\frac{1}{2}\%$, $\frac{1}{2}\%$ and 0·05% points, and also those points which have $2\frac{1}{2}\%$, etc., below them. Now the point shown in Fig. 8.2 having $2\frac{1}{2}\%$ area below it can just as well be said to have $97\frac{1}{2}\%$ of area above it—so in Table II it appears as a $97\frac{1}{2}\%$ point. Example 8.4 illustrates how this is used.

Test 8.4   A sample of N observations is given, from a normal distribution of unknown mean; test whether the distribution could have variance σ^2. Calculate

$$s^2 = \frac{1}{N-1} \sum_{i=1}^{N} (x_i - \bar{x})^2$$

and test $(N-1)s^2/\sigma^2$ as $\chi^2_{(N-1)}$; more directly, calculate $\sum_{i=1}^{N} (x_i - \bar{x})^2$ only, and test $[\sum_{i=1}^{N}(x_i-\bar{x})^2]/\sigma^2$ as $\chi^2_{(N-1)}$. In either case accept the N.H. *variance* $=\sigma^2$ if the value of $\chi^2_{(N-1)}$ falls neither in the lower $2\frac{1}{2}\%$ nor in the upper $2\frac{1}{2}\%$ region of the χ^2 distribution.

EXAMPLE 8.4

The sample of 11 observations 4·3, 1·8, 6·5, 3·2, 5·1, 3·9, 4·6, 4·7, 2·5, 5·8, 3·6 is thought to be taken from a normal distribution with variance 1·21. Test whether this hypothesis is reasonable.

By the usual method, we find $s^2=\frac{1}{10}(19\cdot1764)=1\cdot9176$; thus $19\cdot1764/1\cdot21=15\cdot85$ is $\chi^2_{(10)}$. Now from Table II, the upper $2\frac{1}{2}\%$ point of $\chi^2_{(10)}$ is 20·48, and the lower $2\frac{1}{2}\%$ point, i.e. the upper $97\frac{1}{2}\%$ point in the Table, is 3·25. Our calculated value, being between these, does not require rejection of the N.H., and so we suppose that the given value of the variance is reasonable.

Note 1 If, as is not common, we know the mean, μ, of the distribution, we calculate $\sum_{i=1}^{N}(x_i-\mu)^2$ as the sum of squares, and use N for degrees of freedom in χ^2.

Note 2 If we have a definite Alternative Hypothesis for σ^2, this will require us to make a one-tailed test exactly as already described for means.

Exercises 8.6, 8.7 provide numerical examples of one- and two-tailed tests.

All χ^2-tests except Test 8.4 are not exact, but approximate ones in the sense already described on p. 51. One such test, called the *Index of Dispersion* test, examines whether a set of data follows either the Poisson or the binomial distribution. In Test 8.4, we divided $(N-1)s^2$ by σ^2; if a set of data follows the Poisson distribution, σ^2 should be equal to the mean, of which an estimate $\bar{x}$ is available from the sample. Hence $[(N-1)s^2]/\bar{x}$ is *approximately* $\chi^2_{(N-1)}$, although not exactly so as $\bar{x}$ is only an *estimate* of σ^2. In the same way, σ^2 for a set of binomial data is estimated by $n\hat{p}(1-\hat{p})$, and thus $[(N-1)s^2]/[n\hat{p}(1-\hat{p})]$ is approximately $\chi^2_{(N-1)}$.

Test 8.5a A sample of N observations is given; test the N.H. that it is drawn from a Poisson distribution. Calculate the sample mean, $\bar{x}$, and the sample variance s^2 (or, simply, the sum of squares $\sum_{i=1}^{N}(x_i-\bar{x})^2$, which is $(N-1)s^2$), and test

$$\frac{\sum_{i=1}^{N}(x_i-\bar{x})^2}{\bar{x}} = \frac{(N-1)s^2}{\bar{x}}$$

as $\chi^2_{(N-1)}$.

Test 8.5b   A sample of N observations is given; test the N.H. that it is drawn from a binomial distribution, one of whose parameters is given to be n. Calculate the sample mean $\bar{x}$, and use it to find the estimate $\hat{p}$ of the other parameter (as in Chapter 4); calculate also s^2 (or the sum of squares) and test

$$\frac{\sum_{i=1}^{N}(x_i-\bar{x})^2}{n\hat{p}(1-\hat{p})} = \frac{(N-1)s^2}{n\hat{p}(1-\hat{p})}$$

as $\chi^2_{(N-1)}$.

EXAMPLE 8.5a

Test whether the data of Examples 5.2 and 2.3 do follow the Poisson distribution. We found the mean of the $N = 370$ observations to be $\bar{x} = 4{\cdot}59$, and the variance $s^2 = 4{\cdot}41$ (i.e. the sum of squares is $369 \times 4{\cdot}41 = 1627{\cdot}29$). Thus $\chi^2_{(369)} = 1627{\cdot}29/4{\cdot}59 = 354{\cdot}53$. For very large d.f., as often arise in Index of Dispersion tests, we do not have exact tables, but must use the approximation mentioned beneath Table II: the value of $\sqrt{2\chi^2_{(f)}}$ is tested as $\mathcal{N}(\sqrt{2f-1},\,1)$. In this case, $f = 369$, χ^2 was $354{\cdot}53$, so that $\sqrt{2\chi^2} = \sqrt{2 \times 354{\cdot}53} = \sqrt{709{\cdot}06} = 26{\cdot}63$, and we must test whether it is a member of the normal population $\mathcal{N}(\sqrt{(2 \times 369)-1},\,1)$, i.e. $\mathcal{N}(\sqrt{737},\,1)$ or $\mathcal{N}(27{\cdot}15,\,1)$. The method of Test 7.2 then gives $d = (26{\cdot}63 - 27{\cdot}15)/1 = -0{\cdot}52$, which is certainly not significant; hence we accept that the data follow the Poisson distribution. (Of course, if we had had a much smaller sample, so that the d.f. were in the range of Table II, this normal approximation to the χ^2-test would not be required; but in order to establish what distribution a set of data may follow, a fairly large sample is needed—much larger than for examining the value of a parameter.)

EXAMPLE 8.5b

The numbers of males, r, in 100 separate litters of rats, each litter of total size $n = 5$, were

	$r_i =$	0	1	2	3	4	5	(Exercise 2.1)
	$f_i =$	3	13	30	33	17	4	

In Chapter 3 we found that $\bar{r} = 2{\cdot}6$, $s^2 = 1{\cdot}2727$; and the sum-of-squares $99s^2$ is therefore $126{\cdot}00$. The number of litters examined, 100, is N, and must not be confused with the parameter n, litter size, which is 5. In Chapter 4, the estimate, $\hat{p}$, of the other parameter was found to be $0{\cdot}52$. Let us now test whether this set of data does follow the binomial distribution. If so, $126{\cdot}00/(5 \times 0{\cdot}52 \times 0{\cdot}48)$ is $\chi^2_{(99)}$; so

$\chi^2_{(99)} = 126 \cdot 00 / 1 \cdot 248 = 100 \cdot 96$, certainly not significant and so we can accept the binomial hypothesis.

Comparing two proportions

A common type of statement to be found in reports and in advertising runs like this: 'Our new, improved, strain of seed gives 90% germination, as compared with 80% which is normal for seed of this plant.' If, on careful examination of the experimental conditions, we feel that a fair test has been made, and no factors left uncontrolled which could bias the results, we shall then wish to compare statistically these two percentages or proportions, taking as Null Hypothesis that their difference is 0. A sufficiently significant departure from this N.H. would help to justify the claim.

In Example 7.6, we noted that an estimated proportion, $\hat{p}$, had variance $\hat{p}\hat{q}/N$, and was approximately normal in large samples. Hence the variance of the difference between two estimated proportions, $\hat{p}_1$ and $\hat{p}_2$, is $\hat{p}_1\hat{q}_1/N_1 + \hat{p}_2\hat{q}_2/N_2$; and if N_1, N_2 are large the expression

$$\frac{\hat{p}_1 - \hat{p}_2}{\sqrt{\dfrac{\hat{p}_1\hat{q}_1}{N_1} + \dfrac{\hat{p}_2\hat{q}_2}{N_2}}}$$

is approximately $\mathcal{N}(0, 1)$. This is a perfectly valid test, but rather cumbersome, and another method, also approximate but equally good, is based on the $\chi^2_{(1)}$-distribution.

Test 8.6   In a sample of $(a+b)$ items, a of them possess a certain characteristic and b do not; in a second sample of $(c+d)$ items, c possess the characteristic and d do not. To test the N.H. that the proportions $a/(a+b)$ and $c/(c+d)$ are not significantly different, arrange the data in a table:

Observed nos.	With characteristic	Without characteristic	Total
Sample 1	a	b	$(a+b)$
Sample 2	c	d	$(c+d)$
	$a+c$	$b+d$	N

$N = a+b+c+d$. Test as $\chi^2_{(1)}$:

$$\frac{N(ad-bc)^2}{(a+b)(a+c)(b+d)(c+d)}$$

using a one-tailed test.

EXAMPLE 8.6

If the *special characteristic* is germination, and 100 plants of a new strain gave 90 which germinated (90%) while 200 of an old strain gave 160 which germinated (80%), the table of observed numbers is:

	Germinated	Not germinated	Total
New strain	90	10	100
Old strain	160	40	200
	250	50	300

$$ad-bc = (90 \times 40) - (10 \times 160) = 3600 - 1600 = 2000$$

Thus

$$\chi^2_{(1)} = \frac{300 \times 2000 \times 2000}{100 \times 200 \times 250 \times 50} = 4 \cdot 80$$

greater than the 5% point (which is 3·84) but not so great as the 1% point (6·64). Hence we reject, at the 5% level, the N.H. that the germination rates do not differ significantly, and thus accept that the new strain is better.

Note 1 A one-tailed test is used because very small values of χ^2 would indicate close agreement between the two germination rates, and should certainly not lead us to reject the N.H.; only *large* χ^2 indicates disagreement.

Note 2 The result of the test depends not only on the proportions compared but on the number of observations available; this number, N, has to be much larger than the sample sizes required in *t*-tests, because the information supplied by each observation is less detailed, being only a quality, not an exact measurement.

Note 3 If the table contains any entries that are much smaller than 5, *Yates' correction* may improve the quality of the approximation upon which the test is based. This correction reduces the value of $(ad-bc)$ numerically by $\frac{1}{2}N$ before squaring, so that

$$\chi^2_{(1)} = \frac{N(|ad-bc| - \frac{1}{2}N)^2}{(a+b)(c+d)(a+c)(b+d)}$$

Goodness-of-fit

When qualitative data have been collected, and classified into more than two groups, there may be some hypothesis stating how the numbers in each group relate to one another. In genetic work, an obvious case is the hypothesis that the ratios $AB:Ab:aB:ab$ are $9:3:3:1$. A set of real data is most unlikely to follow the ratios *exactly*, but their goodness-of-fit to the hypothesis can be tested using χ^2.

Test 8.7   A sample of N observations is given, each one of which can be put into one of r categories; the frequencies observed in each category are $O_1, O_2, \ldots, O_r$. Calculate how the frequencies would be expected to fall into the categories upon the given hypothesis; call these frequencies $E_1, E_2, \ldots, E_r$. Then

$$\chi^2_{(r-1)} = \sum_{i=1}^{r} \frac{(O_i - E_i)^2}{E_i}$$

EXAMPLE 8.7

A set of data is expected to show the ratio $9:3:3:1$. A sample of 556 observations gave totals 315, 101, 108 and 32 respectively in the four groups. Test whether this agrees with the given ratio.

If the given ratio operates, $\frac{9}{16}$ of the total number of observations should fall in the first group, $\frac{3}{16}$ in the second, $\frac{3}{16}$ in the third, $\frac{1}{16}$ in the fourth: these *expected* numbers must add to the same total as the observed frequencies.

Construct the table:

Group	1	2	3	4	Total
Observed frequencies	315	101	108	32	556
Expected frequencies	312·75	104·25	104·25	34·75	556

Thus

$$\chi^2_{(3)} = \frac{(315 - 312 \cdot 75)^2}{312 \cdot 75} + \frac{(101 - 104 \cdot 25)^2}{104 \cdot 25} + \frac{(108 - 104 \cdot 25)^2}{104 \cdot 25} + \frac{(32 - 34 \cdot 75)^2}{34 \cdot 75}$$

$$= \frac{2 \cdot 25^2}{312 \cdot 75} + \frac{3 \cdot 25^2}{104 \cdot 25} + \frac{3 \cdot 75^2}{104 \cdot 25} + \frac{2 \cdot 75^2}{34 \cdot 75} = 0 \cdot 47$$

As in Example 8.6, a one-tailed test is used because very small values of χ^2 indicate a very close agreement with hypothesis: the value 0·47 in this example is of this sort. Only if χ^2 had been significantly large (greater than 7·81 for 5% significance with 3 d.f.) could we have rejected the ratios in the hypothesis.

Contingency tables

When each member of a population has been examined for two characteristics, and each characteristic classified into a number of categories, we may want to know if the two characteristics are independent. For example, if each of the plants in a long row is examined for leaf colour (from green to yellow) and vigour (good to weak), it may seem likely that plants with yellow leaves are less vigorous. We can set up a N.H. that leaf colour and vigour are *independent*, and attempt to discredit this.

Test 8.8   Suppose that two characteristics A, B are examined on several individuals. A can take any of the forms A_1, A_2, A_3, ..., A_r and B any of B_1, B_2, ..., B_c. The number of individuals for which A is type A_i and B is type B_j is denoted by the symbol O_{ij}. A table of these observed numbers is constructed, and the row and column totals a_i

Char. A	B_1	Char. B B_2	B_c	Row totals
A_1	O_{11}	O_{12} $\cdots$	O_{1c}	a_1
A_2	O_{21}	O_{22} $\cdots$	O_{2c}	a_2
A_3	O_{31}	O_{32} $\cdots$	O_{3c}	a_3
$\vdots$	.	.	.	$\vdots$
A_r	O_{r1}	O_{r2} $\cdots$	O_{rc}	a_r
Column totals	b_1	b_2	b_c	N

($i = 1$ to r) and b_j ($j = 1$ to c) found, as well as the grand total N. This table is called an $r \times c$ contingency table (it has r rows and c columns). If A and B are independent, the number of individuals to be expected in (A_i, B_j) is $a_i b_j / N = E_{ij}$. Test

$$\sum_{\text{all } i} \sum_{\text{all } j} \frac{(O_{ij} - E_{ij})^2}{E_{ij}} \text{ as } \chi^2_{(r-1)(c-1)}$$

If this is significant, reject the N.H. that A, B are independent.

The formula for E_{ij} arises in the following way. If A, B are independent, then the ratios of $B_1 : B_2 : B_3 : ... : B_c$ should be the same on any A row, and hence the same in the totals $b_1 : b_2 : b_3 ... : b_c$; thus in row A_1, a proportion b_1/N of a_1 should be in the first column, b_2/N of a_1 in the second, and so on.

EXAMPLE 8.8

A sample of 250 seedlings is classified for vigour and leaf colour, with the results below. Test whether these two characteristics are independent.

Observed results:

Leaf colour	Vigour			
	Good	Average	Weak	Total
Green	55	79	4	138
Yellow-green	11	60	15	86
Yellow	1	6	19	26
	67	145	38	250

The expected entries in this table, if colour and vigour are independent, are: Green/Good, $(138 \times 67)/250 = 37{\cdot}0$; Green/Average, $(138 \times 145)/250 = 80{\cdot}0$; Green/Weak $(138 \times 38)/250 = 21{\cdot}0$; and so on, giving the Expected table:

	Good	Average	Weak	Total
Green	37·0	80·0	21·0	138
Yellow-green	23·0	49·9	13·1	86
Yellow	7·0	15·1	3·9	26
	67	145	38	250

Now $(r-1)(c-1) = 2 \times 2$, so that χ^2 has 4 d.f. For Green/Good, $(\text{Obs} - \text{Exp})^2/\text{Exp} = (55 - 37{\cdot}0)^2/37{\cdot}0 = (18{\cdot}0)^2/37{\cdot}0$; for Green/Average, $(79 - 80{\cdot}0)^2/80{\cdot}0$, etc. Adding all nine terms from all nine cells in the table, we obtain

$$\chi^2_{(4)} = \frac{(18{\cdot}0)^2}{37{\cdot}0} + \frac{(1{\cdot}0)^2}{80{\cdot}0} + \frac{(17{\cdot}0)^2}{21{\cdot}0} + \frac{(12{\cdot}0)^2}{23{\cdot}0}$$
$$+ \frac{(10{\cdot}1)^2}{49{\cdot}9} + \frac{(1{\cdot}9)^2}{13{\cdot}1} + \frac{(6{\cdot}0)^2}{7{\cdot}0} + \frac{(9{\cdot}1)^2}{15{\cdot}1} + \frac{(15{\cdot}1)^2}{3{\cdot}9} = 100{\cdot}2$$

This is significant at even more than the $0{\cdot}1\%$ level, so we cannot accept the hypothesis that the colour and vigour are independent; inspection of the tables reveals that there are far more yellow/weak plants observed than would be expected.

Degrees of freedom in Tests 8.7, 8.8 (and 8.6)

In the goodness-of-fit test (Test 8.7), one *constraint* is applied to the expected frequencies, namely that they must add up to the same total as the observed ones, so that when $(r-1)$ of them have been calculated according to hypothesis the last one is automatically determined—it has no *freedom*. In contingency tables (Test 8.8), the totals of expected frequencies in each row must be the same as the totals of observed ones in that row, so that when $(c-1)$ of them are determined, the one in the last column follows automatically; and this goes for $(r-1)$ of the rows, the entries in the last row then following because each column of expected figures must add up to the same total as do the observed figures in that column. So only $(c-1)$ of the entries in each of the first $(r-1)$ rows have freedom. Test 8.6 may, of course, be regarded as a contingency table with 2 rows and 2 columns (a 2×2 Table), and so has 1 d.f.; it may be analysed the same way as a general contingency table if desired.

COMPARING TWO VARIANCES: THE F DISTRIBUTION

Suppose that two samples, of N_1 and N_2 observations respectively, are drawn from normal distributions, and the sample variances s_1^2, s_2^2 calculated; nothing need be known about the means in order to do this (they are, of course, also estimated from the samples). The variance-ratio, that is the ratio of the two variances estimated from the samples, s_1^2/s_2^2, follows the distribution known as F. This distribution is very similar in shape to that of χ^2, covering the range of values 0 to ∞ (positive only), and being skew with a long tail on the right (as is χ^2 in Fig. 8.2); but since it depends on two variable items, s_1^2 and s_2^2, it will have two parameters, namely the degrees of freedom (N_1-1) and (N_2-1). This causes difficulty in preparing tables, and the course usually adopted (as in Table III) is to present, for only a few chosen levels of probability, the values of F which are exceeded with the given probability.

If we set up the N.H. that the samples came from populations with the same variance (but not necessarily the same mean), the expected value of $F = s_1^2/s_2^2$ is 1. In the most common application of F, the Analysis of Variance (Chapters 11, 12, 13), an A.H. is specified which demands a one-tailed significance test; if s_1^2 is not the same as s_2^2 it will have to be greater and cannot be less except through sampling variation. For this reason, F-tables, like χ^2-tables, are always presented in one-tailed form. But in the direct use of F for comparing two sample-variances, a two-tailed test is needed, so that besides the commonly quoted 5%, 1% and 0·1% points, Table III quotes 2½% points.

The ratio

$$s_1^2/s_2^2 \text{ is } F_{[(N_1-1),\ (N_2-1)]} \text{ and } s_2^2/s_1^2 \text{ is } F_{[(N_2-1),\ (N_1-1)]}$$

In testing, it is not necessary for the larger d.f. to be placed first; but what *is* necessary is to arrange that the numerical value of F shall not be less than 1, or in other words the larger s^2 is divided by the smaller. To see if the result of this is significant at one of the tabulated levels, e.g. 5%, look along the top row of Table III to find the first d.f., then down that column until the row corresponding to the second d.f. is reached; the entry at that point in the table is the F-value, with that pair of d.f., that just achieves 5% significance. For example, $F_{(8, 12)} = 2 \cdot 85$, $F_{(1, 24)} = 4 \cdot 26$, $F_{(9, 3)} = 8 \cdot 81$ and $F_{(15, 5)} = 4 \cdot 62$ at 5%.

Test 8.9   Two samples are given, of sizes N_1, N_2 from normal distributions of unknown means. Test the N.H. that the distributions have the same variance. Calculate s_1^2 (with $(N_1 - 1)$ d.f.) and s_2^2 (with $(N_2 - 1)$ d.f.). Suppose $s_1^2 > s_2^2$. Set $F_{[(N_1-1),\ (N_2-1)]} = s_1^2/s_2^2$ and reject the N.H. only if F is significantly large, using a two-tailed test (i.e. upper $2\frac{1}{2}\%$ etc., not 5% etc., points).

EXAMPLE 8.9

Two samples have been drawn at random from normal distributions; for the first, of $N_1 = 15$ observations, the mean $\bar{x}_1$ is $6 \cdot 35$ and the variance s_1^2 is $17 \cdot 2074$, while for the second sample $N_2 = 20$, $\bar{x}_2 = 4 \cdot 18$, $s_2^2 = 7 \cdot 0475$. Test whether the samples could have come from populations with the same variance.

Note first of all that the means are rather different, and could even be significantly so, but this does not affect the test. The ratio $s_1^2/s_2^2 = 17 \cdot 2074/7 \cdot 0475 = 2 \cdot 44$; this is distributed as $F_{(14, 19)}$. The N.H. for this problem is that the variances are equal, and it is being tested only against the vague alternative of inequality, no direction for the inequality being specified; this demands a *two*-tailed test of significance and we have already noted that Table III contains those values of F that are used in *one*-tailed tests. All that we ask of our F-ratio here is that it shall not exceed the upper $2\frac{1}{2}\%$ point (nor be less than the lower $2\frac{1}{2}\%$ point, but this cannot occur since we have arranged that F shall be greater than 1). The upper $2\frac{1}{2}\%$ point for F with 14 and 19 d.f. is $2 \cdot 65$; our value is less than this and so not significant, allowing us to accept that the two populations did have the same variance.

(Had we obtained the same numerical value, $2 \cdot 44$, for $F_{(14, 19)}$ in an Analysis of Variance where a one-tailed test is needed, it would have to be compared with the 5% point in Table III; this is $2 \cdot 26$, and our value exceeds it, so that a one-tailed test would have rejected the N.H.)

EXERCISES

(For Answers and Comments, see p. 167)

8.1 Eight observations were selected at random from a normal distribution, and their values were $1\cdot6$, $-0\cdot8$, $0\cdot1$, -0.4, $1\cdot2$, $0\cdot7$, $0\cdot3$, $0\cdot5$. Test the hypothesis that the normal distribution had mean $0\cdot1$.

8.2 A random sample of 25 seeds is selected from a large population of a certain variety and planted in pots. The mean time from planting to the opening of the first leaflet is $5\cdot85$ days, with variance $4\cdot84$. Assuming that this time follows a normal distribution, test the hypothesis that its mean is 4.

8.3 Two samples of observations on the diameters of mycelial colonies produced the following results. Sample A, of 11 observations, gave a mean value of $6\cdot65$ and a variance of $15\cdot2824$; sample B, of 16 observations, gave a mean value of $4\cdot28$ and a variance of $8\cdot0275$. Find a pooled estimate of variance, and use it in testing whether the samples could have come from distributions with the same mean. (Assume that the distributions of diameters are normal.)

8.4 Soya-bean seedlings were grown in pairs of adjacent pots, one pot of each pair being watered twice as often as the other (though with half the volume of water each time). The differences in height at the end of a period of such treatment, expressed as more regular watering minus less regular in each pair, were $6\cdot0$, $1\cdot3$, $3\cdot1$, $6\cdot8$, $-1\cdot5$, $4\cdot2$, $-3\cdot3$, $2\cdot7$, $10\cdot2$, $0\cdot1$, $-0\cdot4$ mm. Test if there was a significant difference in height due to regularity of watering.

8.5 Experience has shown that the increase in weight of guinea-pigs from a specially bred population, when fed for a standard length of time from birth on a standard diet, is normally distributed with variance 15 units. Six newborn guinea-pigs are selected at random, and fed for the same length of time on a new diet; their weight increases are 18, 21, 12, 16, 25, 20 units. Are these figures significantly different in variability from those obtained on the standard diet?

8.6 Repeat Exercise 8.5 on the assumption that any change in variability due to the new diet is expected to be an increase.

8.7 Calculate an Index of Dispersion to test whether the data of Exercise 5.3 (p. 36) follow a Poisson distribution, and hence whether it appears that the plants are growing at random in the area sampled.

8.8 Calculate an Index of Dispersion to test whether the data of Exercise 4.4 (p. 30) follow a binomial distribution.

8.9 Two preparations of a drug, presented in the same tablet form, are tested for their efficacy in alleviating headaches. Preparation A is given to 250 patients, and 172 of them claim that it is effective; B is given to 200 patients and 158 of these claim that it is effective. Does this provide evidence of a difference between A and B?

8.10 If, under the conditions of Exercise 8.9, A had been given to 25 patients, 17 claiming it effective, while B had been given to 20 patients,

16 claiming it effective, repeat the test for the existence of difference between A and B.

Comment on any apparent discrepancy between the results of the tests in Exercises 8.9 and 8.10.

8.11 One page in a table of random numbers contains 800 digits, and on this page the frequencies of the digits 0, 1, 2, ..., 9 are counted.

Digit	0	1	2	3	4	5	6	7	8	9	*Total*
Frequency	85	77	83	90	69	79	80	76	84	77	800

Do these results contradict the hypothesis that each digit is equally likely to occur at any entry in the table?

8.12 Three strains of corn are thought to be the same genetically. A number of seedlings of each strain are grown, and classified as type a, b, c or d in respect of a certain characteristic. The results are:

Strain	Type				Total examined
	a	b	c	d	
I	75	15	25	5	120
II	85	37	26	12	160
III	60	28	19	13	120

Test whether the strains are the same in respect of this characteristic. Also test whether the ratio of a to $(b+c+d)$ in Strain II could be 9:7.

8.13 The height to which seedlings of two varieties of cider apple grow under standard conditions in a greenhouse is assumed to be normally distributed. Ten seedlings from one variety grow respectively 44, 26, 1, 79, 53, 38, 62, 80, 33, 13 cm, and twelve from the other variety grow 33, 47, 55, 39, 24, 61, 38, 12, 26, 64, 52, 51 cm. Test the hypothesis that the variance of height is the same in the two populations.

8.14 A sample of 9 plants of the same variety were grown in one type of soil in a greenhouse, and after a fixed time they were removed and dried. Their dry weights were 25·5, 22·3, 24·7, 28·1, 26·5, 19·0, 31·0, 25·3, 29·6 g. A further sample of 11 similar plants were grown in identical conditions but in another type of soil. Their dry weights were 31·8, 30·3, 26·4, 24·2, 27·8, 29·1, 25·5, 28·9, 30·0, 26·9, 29·7 g. Do the two soil types have different effects on the plants?

9

Setting Limits to Estimates

A sample of N observations $\{x_1, x_2, x_3, \ldots, x_N\}$ has been used in Chapters 7 and 8 to test various hypotheses about the distribution of x; and when two samples were available, we compared in them the values of suitable parameters such as the mean and the variance. Often it is possible on theoretical grounds to say that the sample has been drawn from a distribution of given type or family, such as normal, binomial or Poisson. We have seen in Chapters 4, 5 and 6 that a complete summary of such a distribution is provided by estimating suitable parameters: μ and σ^2 for the normal distribution, n and p for the binomial, and λ for the Poisson. Of course, since we have only a sample, and that often a small one, upon which to base these estimates, they are attended by some degree of uncertainty. The extent of this uncertainty is important: if it is small, a satisfactory estimate has been made, but if it is large the estimate has little use. Some of the statements of Chapter 8 can be rewritten in a manner which shows the extent of uncertainty in the estimates made, and provides an interval (that is, upper and lower limits) which is very likely to contain the *true* value of the parameter being estimated.

When a sample has been drawn from a normal distribution, the expression

$$t_{(N-1)} = \pm \frac{\bar{x} - \mu}{\sqrt{s^2/N}} \tag{1}$$

tests whether $\bar{x}$ is consistent with some particular value of μ that has been specified. In fact, $\bar{x}$ is consistent with *any* value of μ which keeps the right-hand side of (1) numerically less than $t_{(N-1,\,0.05)}$. (We use this new symbol to denote that value of $t_{(N-1)}$ which, as shown in Table I, just reaches significance at the 5% level.) On multiplying both sides of expression (1) by $\sqrt{s^2/N}$, we see that the sample could arise from any normal distribution whose mean μ has a value that makes $\pm(\bar{x}-\mu)$ not greater than the product of $t_{(N-1,\,0.05)}$ and $\sqrt{s^2/N}$. Writing this out separately for the positive and negative signs, we must have

$$+(\bar{x}-\mu) \leqslant t_{(N-1,\,0.05)}\sqrt{\frac{s^2}{N}} \quad \text{and also} \quad -(\bar{x}-\mu) \leqslant t_{(N-1,\,0.05)}\sqrt{\frac{s^2}{N}}$$

so that

$$\bar{x}-t_{(N-1,\,0.05)}\sqrt{\frac{s^2}{N}} \leqslant \mu \quad \text{and also} \quad \mu \leqslant \bar{x}+t_{(N-1,\,0.05)}\sqrt{\frac{s^2}{N}}$$

Combining these two statements, we have

$$\bar{x}-t_{(N-1,\,0.05)}\sqrt{\frac{s^2}{N}} \leqslant \mu \leqslant \bar{x}+t_{(N-1,\,0.05)}\sqrt{\frac{s^2}{N}} \qquad \textbf{(A)}$$

Since **(A)** contains a value of $t_{(N-1)}$ which will be exceeded by random chance on 5% of occasions even when the sample mean *is* consistent with μ, **(A)** cannot be an absolutely true statement; but it provides a range of values that contain the true μ with 95% probability, on the evidence supplied by the sample (that is to say, the observed values of $\bar{x}$, s^2 and N). In other words, if we make the assertion **(A)** we can have 95% confidence that it gives a range of values containing the true μ; we call **(A)** 95% *confidence limits* for μ. The arithmetic of calculating confidence limits is quite similar to that of hypothesis testing; but the two ideas are really very different. Confidence limits use a sample of observations to give us information that we did not have before taking the sample, about the mean or some other parameter in a distribution; for a hypothesis test we do have information, an idea or a theory, which the sample is used to check.

EXAMPLE 9.1 (data of Example 8.1)

A sample of 10 observations drawn at random from a normal population has mean 1·58 and variance 1·513. Set 95% confidence limits to the true value, μ, of the mean of the population.

First calculate

$$t_{(9,\,0.05)}\sqrt{\frac{1\cdot513}{10}} = 2\cdot26\sqrt{0\cdot1513} = 2\cdot26\times0\cdot389 = 0\cdot88$$

The required limits, by **(A)**, are then $1\cdot58-0\cdot88 \leqslant \mu \leqslant 1\cdot58+0\cdot88$ or $0\cdot70 \leqslant \mu \leqslant 2\cdot46$. So, with 95% confidence of being correct, we may say that the mean of the parent normal population, from which the sample was drawn, was not less than 0·70 nor greater than 2·46.

Note that the two-tailed value of *t* must always be used, because we quote upper and lower limits to μ.

LIMITS USING THE NORMAL DISTRIBUTION

If our sample is drawn from a normal distribution in which the value of σ^2 is known, we shall obviously use this σ^2 in calculating limits for μ. In this case there is no need to introduce the *t* distribution, and we may base calculations on the expression used in Test 7.3 rather than that of Test 8.1. The effect of this is to alter (**A**) by writing σ^2 instead of s^2, and $d_{(0.05)}$ instead of $t_{(N-1,\ 0.05)}$, so obtaining

$$\bar{x} - d_{(0.05)}\sqrt{\frac{\sigma^2}{N}} \leqslant \mu \leqslant \bar{x} + d_{(0.05)}\sqrt{\frac{\sigma^2}{N}}$$

In practice, it is unlikely that σ^2 in a normal distribution would be known while μ was not, so this result is not often applied directly. But it leads to a valuable method of dealing with parameters of other distributions when (by virtue of the Central Limit theorem) they are approximately normally distributed.

When estimating the value of *p* in a binomial distribution, we calculate $\hat{p}$; we have already noted (p. 53) that when the sample is a large one and *p* not too near 0 or 1, $\hat{p}$ is approximately $\mathcal{N}(p, \hat{p}\hat{q}/N)$. So in the expression given above, μ is replaced by *p*, $\bar{x}$ by $\hat{p}$ and σ^2/N by $\hat{p}\hat{q}/N$. Hence we obtain the expression

$$\hat{p} - d_{(0.05)}\sqrt{\frac{\hat{p}\hat{q}}{N}} \leqslant p \leqslant \hat{p} + d_{(0.05)}\sqrt{\frac{\hat{p}\hat{q}}{N}} \qquad \textbf{(B)}$$

giving approximate 95% limits to *p*. As the limits are in any case approximate (because $\hat{p}$ does not follow the normal distribution exactly) it is common to replace $d_{(0.05)}$, whose true value is of course 1·96, by the factor 2 in expression (**B**) for 95% limits; this simplifies the arithmetic without seriously affecting the approximation.

EXAMPLE 9.2

Of 250 insects treated with a certain insecticide, 180 were killed. Set approximate 95% confidence-limits to the value of *p*, the proportion of insects likely to be killed by this insecticide in future use.

Here

$$N = 250 \quad \text{and} \quad \hat{p} = \frac{180}{250} = 0 \cdot 720$$

Thus

$$\frac{\hat{p}\hat{q}}{N} = \frac{0 \cdot 72 \times 0 \cdot 28}{250} = \frac{0 \cdot 2016}{250} = 0 \cdot 0008064$$

Therefore

$$2\sqrt{\frac{\hat{p}\hat{q}}{N}} = 2 \times \sqrt{0 \cdot 0008064} = 2 \times 0 \cdot 0284 = 0 \cdot 057$$

So approximate 95% limits for p are $0 \cdot 720 - 0 \cdot 057 \leqslant p \leqslant 0 \cdot 720 + 0 \cdot 057$, i.e. p is between $0 \cdot 663$ and $0 \cdot 777$ (with 95% confidence in the truth of this statement).

Such relatively wide limits for p (percentage kill between $66 \cdot 3$ and $77 \cdot 7$) may not be good enough for some purposes; note that by increasing sample size by a factor k^2 we decrease the width of the limits by a factor k (for if N becomes $k^2 N$, then $\sqrt{\hat{p}\hat{q}/N}$ becomes $\sqrt{\hat{p}\hat{q}/k^2 N}$ or $(1/k)\sqrt{\hat{p}\hat{q}/N}$). In this example a sample of 1000, of which 720 were killed, would provide approximate 95% limits of $0 \cdot 692$ to $0 \cdot 748$ for p; by increasing sample size four-fold we halve the width of the limits.

The mean and variance of a Poisson distribution are both equal to λ (Chapter 5). Therefore, from a sample of N observations, λ is estimated by the sample mean $\bar{x}$; and the variance of this sample mean will be estimated by $\bar{x}/N$. Provided the sample size, N, is not too small, $\bar{x}$ will be approximately normally distributed, and in this case approximate 95% limits will be

$$\bar{x} - 2\sqrt{\frac{\bar{x}}{N}} \leqslant \lambda \leqslant \bar{x} + 2\sqrt{\frac{\bar{x}}{N}} \qquad \text{(C)}$$

EXAMPLE 9.3

A sample of 121 observations from a Poisson distribution had a mean of $11 \cdot 0$. Set approximate 95% limits to the parameter λ of the distribution.

Now

$$\sqrt{\frac{\bar{x}}{N}} = \sqrt{\frac{11 \cdot 0}{121}} = \sqrt{\frac{1}{11}} = 0 \cdot 302$$

so $2\sqrt{\bar{x}/N} = 0 \cdot 60$, giving as approximate 95% limits $11 \cdot 0 - 0 \cdot 60 \leqslant \lambda \leqslant 11 \cdot 0 + 0 \cdot 60$; thus λ is between $10 \cdot 4$ and $11 \cdot 6$ (with 95% confidence).

When we have a very large sample of observations, we may calculate limits to the mean μ of the distribution from which the sample was drawn, even if we do not know that the distribution is normal. This is because the sample mean $\bar{x}$ will itself follow an approximately normal distribution in a large sample, in accordance with the Central Limit Theorem. When **(A)** is used for very large samples, it becomes approximately

$$\bar{x} - d\sqrt{\frac{s^2}{N}} \leqslant \mu \leqslant \bar{x} + d\sqrt{\frac{s^2}{N}}$$

Here μ is estimated by $\bar{x}$, whose standard deviation, or *standard error* as it is commonly called, is $\sqrt{s^2/N}$, and d is usually set equal to 2 rather than 1·96: so limits for the parameter (μ) are its estimate ($\bar{x}$) plus or minus twice the standard error of the estimate. This phrase *estimate plus or minus twice its standard error* is very often used to indicate how to calculate approximate 95% limits to parameters whose estimates have nearly normal distributions. (We usually speak of the standard *deviation* of a single observation from a parent distribution, but of a standard *error* of some function, such as a mean or a variance, which has been calculated from a sample: otherwise there is no difference in meaning between the two phrases.)

The confidence level at which intervals are calculated is most often 95%, but if desired can be 99% or 99·9%. In **(A)**, $t_{(N-1, \, 0.05)}$ would be replaced by $t_{(N-1, \, 0.01)}$ or $t_{(N-1, \, 0.001)}$; elsewhere, d (or 2) would be replaced by 2·58 or 3·29.

LIMITS TO VARIANCES

When a sample of N observations was available from a normal distribution, and the sample variance s^2 was calculated, we used in Test 8.4 the expression $[(N-1)s^2]/\sigma^2$. If this expression lay between the upper and lower $2\frac{1}{2}\%$ points of $\chi^2_{(N-1)}$, then the sample variance s^2 was consistent with a hypothetical value of σ^2 for the true variance in the population. Call the upper and lower $2\frac{1}{2}\%$ points, respectively, $\chi^2_{(N-1, \, 0.025)}$ and $\chi^2_{(N-1, \, 0.975)}$ as in Table II. Then, by an argument similar to that used at the beginning of this chapter, limits to the possible values of σ^2, based on the variance s^2 of the sample of N observations, are obtained from

$$\chi^2_{(N-1, \, 0.975)} \leqslant \frac{(N-1)s^2}{\sigma^2} \leqslant \chi^2_{(N-1, \, 0.025)}$$

This is more conveniently written

$$\frac{(N-1)s^2}{\chi^2_{(N-1, \, 0.025)}} \leqslant \sigma^2 \leqslant \frac{(N-1)s^2}{\chi^2_{(N-1, \, 0.975)}} \qquad \textbf{(D)}$$

EXAMPLE 9.4

Using the data of Example 8.4, set 95% confidence limits to σ^2.

Here $(N-1) = 10$, $(N-1)s^2 = 19 \cdot 1764$, $\chi^2_{(10,\ 0 \cdot 025)} = 20 \cdot 48$ and $\chi^2_{(10,\ 0 \cdot 975)} = 3 \cdot 25$. Thus $19 \cdot 1764/20 \cdot 48 \leqslant \sigma^2 \leqslant 19 \cdot 1764/3 \cdot 25$, so that $0 \cdot 936 \leqslant \sigma^2 \leqslant 5 \cdot 90$, giving the range that contains the true value of σ^2, with 95% confidence.

It is also possible, using the F-distribution, to obtain limits to the ratio of two variances; this is a less straightforward task and readers are referred to more advanced books.

EXERCISES

(For Answers and Comments, see p. 169)

9.1 (a) Set 95% and 99% confidence-limits to the mean of the normal distribution from which the sample of 8 observations in Exercise 8.1 (p. 75) was taken.

(b) Carry out similar calculations for the data of Exercise 8.2.

(c) Carry out similar calculations for each of the 2 samples, A and B, of Exercise 8.3.

9.2 Find 95% and 99% confidence-limits for the difference in height of soya-bean seedlings, due to regularity of watering, using the data of Exercise 8.4.

9.3 (a) Find approximate 95% confidence-limits for the proportion of patients cured by each of the preparations A, B described in Exercise 8.9.

(b) Carry out a similar calculation for the ratio of a to $(b + c + d)$ in Strain II of Exercise 8.12.

9.4 How many patients would have to be tested with preparation A (Exercise 8.9) in order to estimate the proportion of cures to within $\pm 3\%$ (at the 95% confidence level)?

9.5 Fifty samples, each of 1 unit volume, were drawn at random from a suspension containing cells. Each sample was placed on a microscope slide and examined; it was found that the average number of cells per unit volume for these 50 samples was 4. Set approximate 95% confidence-limits to the mean number of cells per unit volume in the whole suspension.

9.6 Set 95% confidence-limits to the variances of height in each of the two samples of cider apple seedlings referred to in Exercise 8.13 (p. 76).

9.7 A large normal population has mean 100 and variance 10. Find 95% confidence-limits to

(a) the value of one single observation drawn from this population,
(b) the mean of 10 randomly selected observations drawn from this population,
(c) the mean of 100 randomly selected observations drawn from this population.

9.8 If, in Exercise 9.7, the variance had not been known, how could limits be found?

IO

Correlation

So far we have been concerned entirely with populations wherein one measurement only has been taken on each member; we have called it x_i or r_i. Sometimes measurements might have been taken before and after treatment, but they would be of the same characteristic (e.g. weight of an animal) and most probably the really interesting variate would be the difference between the two. Essentially, one single characteristic of the population has been under study.

There are occasions when two measurements, x_i and y_i, are taken simultaneously on each member of a population, for the purpose of seeing whether they are related. The reason for this may be pure scientific interest, such as measuring the height and weight of each member of an animal population to see if any general laws or patterns of relationship between height and weight can be set up; or it may be the search for measurements y_i, which are easier to take than the x_i that are really of interest, and are at the same time sufficiently closely related to the x_i to give the same information. Thus it is common to measure the amount of a chemical present in a solution, x_i, by the optical density of the solution, y_i, since the latter can be determined at once on a standard laboratory instrument once this has been suitably calibrated (Example 11.1): we do not wish to know y_i, but it is a convenient short cut to x_i, in which we are really interested.

Correlation coefficients used to be calculated almost whenever an opportunity presented itself, but in recent work they have taken a much less prominent place as more detailed methods of examining relationship have been developed: some of these will be studied in later Chapters.

The true (but unknown) value of the correlation coefficient in a population is denoted by ρ, and its estimate, calculated from a sample as described below, by r. Assume that a sample of N members has been chosen at random from a population and examined, two variates x_i and y_i having been recorded for the ith member of the chosen sample. The variances of x and y are found as in Chapter 3, and one further calculation is required: the **covariance** of x and y, in which $(x_i - \bar{x})^2$ is replaced by the product $(x_i - \bar{x})(y_i - \bar{y})$. The formal definition is

$$\operatorname{cov}(x, y) = \frac{1}{N-1} \sum_{i=1}^{N} (x_i - \bar{x})(y_i - \bar{y})$$

and another form for calculation is

$$\frac{1}{N(N-1)} \left[N \sum_{i=1}^{N} (x_i y_i) - G_x G_y \right]$$

The symbols G_x, G_y stand for the two grand totals, and the expression $\sum_{i=1}^{N} x_i y_i$ means 'for each member of the chosen sample, multiply the measurement of x by the measurement of y on that member, and add these products xy for all members of the sample'.

Then $r = \operatorname{cov}(x, y)/s_x s_y$, in which s_x and s_y denote the standard deviations of x and y: although these standard deviations are, by definition, positive, the expression $\operatorname{cov}(x, y)$ need not be. In fact, since r is to measure the relation between x and y, we require it to be positive when x and y increase together and negative when one is increasing as the other decreases. This is exactly what the sum of products $\sum_{i=1}^{N}(x_i - \bar{x}) \times (y_i - \bar{y})$ will achieve, for it will be negative if larger-than-average x_i are nearly always accompanied by smaller-than-average y_i. By dividing $\operatorname{cov}(x, y)$ by $s_x s_y$ we obtain a measure which can genuinely be called a coefficient, since its numerical value cannot be greater than 1.

For calculation of r the divisor $N(N-1)$ can be omitted from numerator and denominator, since it occurs in both; this gives the simplest form

$$r = \frac{N \sum_{i=1}^{N} (x_i y_i) - G_x G_y}{\sqrt{(N \sum_{i=1}^{N} x_i^2 - G_x^2)(N \sum_{i=1}^{N} y_i^2 - G_y^2)}}$$

the $\sqrt{}$ always being taken positive.

Fig. 10.1 illustrates the way in which values of r indicate the relation between x and y; each cross on the diagrams represents a pair of observations (x_i, y_i) measured on an individual. In Fig. 10.1(a) r is +ve and fairly large, perhaps about 0·75; in (b) $r = +1$; in (c) $r = -1$; in (d) r is −ve and fairly large; in (e) and (f) r is approximately 0. We see that a correlation coefficient is really measuring how close the relation between x and y is to *linearity*, for if this relation is perfect as in (b) and (c) r takes its maximum possible numerical value: the actual slope of the line can be found (as in Chapter 11) when we know in what units x, y are measured.

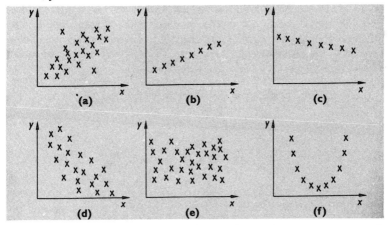

Fig. 10.1 Values of the correlation coefficient, r, between x and y: (a) r positive and fairly large; (b) $r = +1$; (c) $r = -1$; (d) r negative and fairly large; (e) r approximately 0; (f) r approximately 0.

In (a) and (d) the points are scattered about a reasonably well-discerned line, while in (e) it is clear that x may take any value without restricting the value taken by y. But (f) should be a sharp warning against the indiscriminate use of correlation: r is virtually zero in cases like this because the values of one variate, in this example y, are symmetrically distributed about the mean of the other, x. In other words, there are y_i-values of identical size at each of the two points $+(x_i - \bar{x})$ and $-(x_i - \bar{x})$. This symmetrical distribution may be, and often is, a perfectly good, valid and interesting relation between x and y.

The steps in using the correlation coefficient should thus be: (1) plot the data on a graph (this is often called a *scatter diagram*); (2) see if there is any apparently curved relation between the variates; (3) if there is *not* a curve, and particularly when the scatter diagram suggests a *linear* relation, then calculate r. The value of r in these circumstances will give a valid measure of the size of linear relation.

The significance of the calculated value of r may be tested using Table IV, in which the degrees of freedom must be taken as $N - 2$ when N pairs of records (x_i, y_i) are available; r must be at least as large numerically (positive or negative) as the value in the Table before it can be called significant. The Null Hypothesis here is that $\rho = 0$; that is, we test whether the calculated value of r (either $+$ or $-$, the sign being left out of the

Table) is significantly different from zero. If there were reasons to suggest that ρ had some value other than o, i.e. we were testing a calculated value against a hypothesis that $\rho = \frac{1}{2}$, or $\frac{3}{4}$, or $\frac{1}{3}$, etc., Fisher's z-test would be required; see Clarke and Cooke[2] and Fisher[6] for this, and for testing the difference between two correlation coefficients, calculated from two different samples.

Various pitfalls exist in interpreting the significance of correlation coefficients, even assuming it was justifiable to calculate them after looking at the data. These occur when the existence of a correlation is then logically equated to a *causal* relation: x causes y or x is due to y. If we are merely looking at x and y to see if they seem to be showing the same trend, so that we then choose to measure only one of them (the more convenient) in order to demonstrate the trend, the logic of cause-and-effect is not so important. But theories have frequently been built on correlations between x and y when these really represented two measurements each quite closely related to some *third* one, z, and not to one another at all directly. The most picturesque examples exist in the social sciences, when z represents time. For example, if x_i denotes the number of television receiving licences taken out in the ith year since 1945, and y_i the number of convicted juvenile delinquents in the same year, this set of measurements showed a considerable, positive correlation over the twenty years 1945–64. Some people would argue that this meant that television had a bad influence on the young; but in fact a similar result could be found by comparing x_i = number of radio licences and y_i = number of people certified insane, over a fifteen-year pre-war period, and an argument in this case might be harder to sustain! The much more likely answer is that all these things are increasing with time (and population) for, perhaps, not entirely unconnected reasons but certainly not as mutual cause and effect.

EXAMPLE 10.1

A group of $N = 20$ strawberry plants was grown in pots in a greenhouse, and measurements were taken on y, the crop yield, and x, the corresponding level of nitrogen present in the leaf at the time of picking

x: 2·50　2·55　2·54　2·56　2·68　2·55　2·62　2·57　2·63　2·59
y: 247　245　266　277　284　251　275　272　241　265

x: 2·69　2·61　2·67　2·57　2·53　2·70　2·51　2·58　2·53　2·61
y: 281　292　285　274　282　295　249　246　261　260

(x is measured in parts per million by weight of dry leaf matter, and y in g)

$$N = 20, \qquad G_x = 51\cdot79, \qquad G_y = 5348,$$

$$\sum_{i=1}^{20} x_i^2 = 134\cdot1793, \qquad \sum_{i=1}^{20} y_i^2 = 1435404, \qquad \sum_{i=1}^{20} x_i y_i = +13859\cdot60$$

Therefore

$$r = \frac{20 \times 13859\cdot60 - 51\cdot79 \times 5348}{\sqrt{(20 \times 134\cdot1793 - (51\cdot79)^2)(20 \times 1435404 - (5348)^2)}} = +0\cdot5698$$

This has 18 d.f., and is significant at the 1% level (being greater than the value $0\cdot5614$ in Table IV), so that we must reject the Null Hypothesis that $\rho = 0$, and accept that there does seem to be a relation between x and y.

This leads to an illustration of another error in interpreting correlation coefficients. Let us suppose that a further set of 50 plants is now examined, but that the fertilizer applied to them contains nitrogen at a level so high that all are likely to be giving their maximum yield. In this case, any variations in x would not be accompanied by equivalent, predictable variations in y, so that no significant correlation would exist. What result would be obtained by working out a correlation coefficient between x and y for the combination of both samples of plants, that is of the whole 70? It would be dominated by the pattern of behaviour in the larger group, the one of 50 plants; these would swamp the smaller group of 20, and the net result would be that apparently x and y are not significantly correlated over the whole set of 70 plants. But this is fallacious because the 70 are not a homogeneous set: they can be broken down into groups (in this case into just two groups) wherein the growth and cropping features are not the same. Had we plotted all 70 points on a graph, it would have been clear that the same linear relation did not exist throughout: a graph like Fig. 15.1 in shape (though, of course, with many more points) would very likely have been found (p. 134).

If there is no other way of establishing a theory than by using correlation, great care is needed to eliminate all possibility of indirect relationship, that is via a third variable z: the controversy about cigarette smoking and lung cancer shows how much play can be made on this theme. In this particular instance, whatever *third variables* have been suggested, the relation has seemed to remain when they have been eliminated.

There is no straightforward method of setting limits to calculated values of r, particularly for small sample sizes, N, because the distribution of r is not symmetrical enough to use standard-error methods.

The value of r, as a measure of relation between x and y, is a reliable indicator only when x and y are normally distributed, so that it is wise to check that this condition is, at least approximately, satisfied before making the calculation.

RANK CORRELATION

We may not always be able to measure continuous variates x and y on the N members of a population; but instead it may be quite possible to measure an x and a y, each of which can be placed in order for the N members. For example, the judge at the village Flower Show may be faced with 15 pots of strawberry jam that have to be ranked in order 1, 2, 3 and the rest. Important characteristics will include colour and texture (stiff, about right, or runny) and of course flavour: let us concentrate on the first two. Neither of these characteristics can be measured on a continuous scale, but it is very likely that the 15 pots can each be given a position in the rank order 1, 2, . . . for colour (1 is the best, down to 15 which is the worst) and a rank for texture also (1 is the nearest to good texture, down to 15 which is the worst—either extremely stiff or extremely fluid depending on what the contestants have offered!). How well do the rankings for colour agree with those for texture? Is the best for colour the best for texture, or at least are the best four or five for colour also the best four or five for texture; or does there seem to be no relation at all between the two rank orders?

We label the rank order for colour x and that for texture y, and we will assume that the judge has ranked all 15 pots of jam in the following orders.

Pot	a	b	c	d	e	f	g	h	j	k	l	m	n	p	q
Rank for colour, x	10	14	6	1	7	4	9	13	2	12	5	11	15	3	8
Rank for texture, y	4	11	5	6	12	1	14	10	7	13	2	15	9	3	8

In spite of x and y not being continuous measurements, we will calculate the correlation coefficient between them just as though they were. This process is due to Spearman, and the *rank correlation* coefficient so obtained is called Spearman's coefficient, usually denoted by r_s. Of course it was necessary to study the distribution of r_s, which Spearman did, obtaining the results which we shall use later in testing its value.

The easiest way to calculate r_s is first to add a line to the table of records above, giving d_i, the difference in ranks $(x_i - y_i)$ for each member of the sample; these are

Pot	a	b	c	d	e	f	g	h	j	k	l	m	n	p	q
d_i	6	3	1	-5	-5	3	-5	3	-5	-1	3	-4	6	0	0

There is a useful check on the correctness of our calculation: the sum of the d_i should be 0.

It can be shown, after some algebra (see, for example, Clarke and Cooke,[2] section 20.9), that Spearman's r_s is now given by

$$r_s = 1 - \frac{6\sum_{i=1}^{N}d_i^2}{N(N^2 - 1)}$$

in which N is the number of values of d_i available, that is the number of items being ranked. In our set of records, $N = 15$ and so $N(N^2 - 1) = 15 \times 224 = 3360$. Also $\sum d_i^2 = 6^2 + 3^2 + 1^2 + \ldots + 6^2 + 0^2 + 0^2 = 226$, so that

$$r_s = 1 - \frac{6 \times 226}{3360} = 1 - 0 \cdot 4036 = +0 \cdot 5964.$$

We shall now wish to test the significance of the value which we have calculated. Being a correlation coefficient, it has to lie between -1 and $+1$; the value $+1$ will occur when the two rankings x and y agree exactly, and -1 when they disagree completely (the best for colour is worst for texture, the second best for colour is the second worst for texture, and so on right through the list). The Null Hypothesis usually needed is that there is *no relation* between the two rankings x, y; in other words if a sample of jam has been ranked high in the list for colour this tells us nothing at all about what to expect for texture—it may be good, bad or indifferent. As with an ordinary correlation coefficient this implies $r_s = 0$.

For values of N up to 10, a special table is needed (Table V) which shows what value of r_s (positive or negative) is necessary if it is to be significantly different from 0. For larger N, either we may use the table for the ordinary correlation coefficient (Table IV), or to obtain a slightly better approximation we may use the fact that, for $N > 10$, the expression

$$r_s \sqrt{\frac{N-2}{1-r_s^2}}$$

has approximately the t distribution with $(N-2)$ d.f. In our example, therefore, $0 \cdot 5964 \sqrt{\dfrac{13}{1 - 0 \cdot 3557}}$ may be tested as $t_{(13)}$. This is $0 \cdot 5964 \times 4 \cdot 4918 = 2 \cdot 68$, which is significant at the 5% level, leading us to reject the N.H. of no relation between the two rankings. Although some pots of jam have been given very different rankings for the two characteristics, most are either reasonably high for both or quite low for both.

EXERCISES

(For Answers and Comments, see p. 170)

10.1   The blood-clotting times of a number of people were measured before and after taking a certain drink; the times before, x sec, and after, y sec, are recorded below. Calculate the correlation coefficient between x and y.

Subject	A	B	C	D	E	F	G	H	J	K	L	M	N	P	Q
x	175	142	124	168	117	134	167	147	126	104	136	129	178	146	149
y	82	90	126	128	127	54	117	100	91	89	61	134	78	106	99

10.2   Measurements of the heights (inches) of brother and sister were made in each of 15 two-child families, with the following results. Calculate the correlation coefficient between the two heights.

Family	1	2	3	4	5	6	7	8	9	10	11	12	13	14	15
Brother x	73	70	74	68	70	67	69	71	70	68	69	68	71	73	69
Sister y	69	67	63	66	67	64	66	68	67	65	65	64	66	67	67

10.3   'The leaf area of the tomato plants used in the experiment was positively correlated with the height to which the plants grew.' What use could be made of the information in this extract from a report?

10.4   Two measurements x, y are made on each of 40 people; x is scored 0 or 1 for each person and y is scored -1 or $+1$. A correlation coefficient is calculated between x and y and is found to be significantly different from 0. What can be inferred from this?

10.5   Comment on the following statements.

(a) 'The correlation coefficient between x, the mean annual rainfall at an observation station, and d, the station's distance inland, was -0.79 for 38 stations in Sierra Leone. This shows conclusively that rainfall is confined mainly to the coastal strip.'

(b) 'The correlation coefficient between the age, x, of a sow and the number of piglets, y, produced per litter was calculated for all the sows of various breeds on the farm. It was not significantly different from 0, so that (contrary to suspicion) there is no tendency for litter size to decrease with sow's age.'

10.6   The percentage of a certain chemical is estimated, in each of 12 samples A–M, by two methods. Method I is much quicker than Method II, but the chemist doing the estimation suspects that it may give less reliable results: in particular he thinks that the figure 89 for E is wrong, and should have been 59. Calculate a *rank* correlation coefficient between the results on the two methods. What advantage might this have over an ordinary correlation coefficient?

Sample	A	B	C	D	E	F	G	H	J	K	L	M
Method I	46	48	55	47	89	45	44	42	50	63	41	49
Method II	42	44	38	36	58	39	37	45	40	47	43	41

10.7 Two organisms, X and Y, are each grown in eight different environmental conditions I–VIII. Their sizes after a fixed time are (volumes in cm^3):

Environment	I	II	III	IV	V	VI	VII	VIII
Organism X	22	16	38	187	24	68	31	478
Organism Y	44	48	32	155	42	93	35	336

Calculate, and test for significance, a rank correlation coefficient between the sizes of X and Y in the different environments.

The ordinary correlation coefficient is $+0.996$. Comment on the difference between this and the rank coefficient.

II

Linear Regression

When two variates x, y do have a significant correlation coefficient, or when the measurements have been shown on a graph and look like Fig. 10.1(a), (b), (c) or (d), it is natural to try to describe the linear, or approximately linear, relation between them by an equation, the **linear regression** equation. This equation expresses one variate, the *dependent* variate y, in terms of the other, the *independent* variate x. We assume for the moment that it is clear which of the variates is the dependent one, as in Example 11.1 where the optical density, y, of a solution depends on the concentration, x, of the chemical present in it, so that we can properly write $y = a + bx$ to show how y alters when x is altered. Fig. 11.1 indicates what a and b are: a is the *intercept*, or that value taken by y when x is 0, while b is the *slope* or gradient of the line, the average increase in y for unit increase in x.

We can choose, and fix, the values of x to use in an experiment designed to study this relation, but we have no control over the resulting y. The idea, and the logic, behind calculating a regression line is different from that for correlation: in regression, one measurement y is a variable and depends on the other measurement x which can be fixed, while in correlation neither x nor y can be fixed—both are variables.

EXAMPLE 11.1

The optical density, y, of a solution measured at eight concentrations, x, of a chemical was as follows (illustrated in Fig. 11.1):

Meter reading

y_i: 4 9 18 20 35 41 47 60   Total $\sum y_i = 234$

Conc. (μg/ml)

x_i: 1 2 4 5 8 10 12 15   Total $\sum x_i = 57$

We shall find the line which fits the observed points best, in the sense that for each given value of x_i the value of y_i on the line (the *predicted* value of y_i) is as near as possible to the observed value of y_i. Note that because we are treating x as fixed, no question of *fitting* for x arises.

Fig. 11.1 Relation between optical density, *y*, of a solution and concentration, *x*, of chemical present in it (Example 11.1).

Consider the sum of the squares of all the vertical deviations like QP on Fig. 11.1 (Q is the observed value of y at $x = 8$, P is the y-value on the fitted line), i.e. consider $S_D = \sum_{i=1}^{N} (y_{i.\text{Obs}} - y_{i.\text{Pred}})^2$ in an obvious notation. The *least-squares* method of estimating a and b involves putting $y_{i.\text{Pred}} = a + bx_i$ and then looking for those values of a, b which will make S_D as small as possible. In other words, the required estimates $\hat{a}$ and $\hat{b}$, of a and b, will be those which minimize $S_D = \sum_{i=1}^{N} (y_{i.\text{Obs}} - a - bx_i)^2$. When the necessary mathematics has been done, we find that

$$\hat{b} = \frac{\sum_{i=1}^{N} (x_i - \bar{x})(y_i - \bar{y})}{\sum_{i=1}^{N} (x_i - \bar{x})^2}, \quad \text{and} \quad \hat{a} = \bar{y} - \hat{b}\bar{x}$$

so that $y = (\bar{y} - \hat{b}\bar{x}) + \hat{b}x$, or more neatly $y - \bar{y} = \hat{b}(x - \bar{x})$. To calculate b, the same sort of algebra is used as for the correlation coefficient, leading to

$$\hat{b} = \frac{[N \sum_{i=1}^{N} x_i y_i - G_x G_y]}{[N \sum_{i=1}^{N} x_i^2 - G_x^2]}$$

In Example 11.1, $N = 8$, $G_x = 57$, $G_y = 234$. Therefore $\bar{x} = 7.125$, $\bar{y} = 29.25$.

$$\sum_{i=1}^{8} x_i^2 = 1^2 + 2^2 + 4^2 + \cdots + 15^2 = 579$$

$$\sum_{i=1}^{8} x_i y_i = (1 \times 4) + (2 \times 9) + \cdots + (15 \times 60) = 2348$$

$$\therefore \hat{b} = \frac{8 \times 2348 - 57 \times 234}{8 \times 579 - 57^2} = \frac{5446}{1383} = 3\cdot94$$

$\hat{a} = 29\cdot25 - 3\cdot94 \times 7\cdot125 = 1\cdot19$. Therefore the slope of the line is $3\cdot94$, y increasing by this amount for each unit increase in concentration x; and when $x = 0$, y takes the value $1\cdot19$, which in this example could represent some sort of zero- or calibration-error of the instrument. To draw the line on the graph, mark the points $(x = 0, y = a)$ and $(x = \bar{x}, y = \bar{y})$, and join them by the regression line: in this case, the two points are those marked by circles in Fig. 11.1. A regression line always passes through the point $(\bar{x}, \bar{y})$. Of course, b, like ρ, can be negative, if the relation between x and y is as in Fig. 10.1(d). The equation is presented as $y - 29\cdot25 = 3\cdot94(x - 7\cdot125)$, or as $y = 1\cdot19 + 3\cdot94x$. The least-squares technique for finding $\hat{a}$, $\hat{b}$ requires no assumptions about the pattern of variation either in y or in $(y_{Obs} - y_{Pred})$; but clearly we need to measure the goodness-of-fit of the fitted line to the values which were actually observed, and some distributional assumptions are required to do this. So far we have laid down only one condition: *that x should be fixed, capable of observation without error* (1). We now add a second, *that the errors in measurement of y shall be independent of one another and of x, and shall be normally distributed with constant variance, σ^2* (2).

The method of testing goodness-of-fit of a linear regression is based on the **Analysis of Variance** which we shall meet in another connection in Chapter 12. The name arises because we use an expression that is very like a variance (**S** below), and analyse it into its constituent parts, each one of which represents some meaningful portion of the whole. Define the **Total Sum of Squares**, $S = \sum_{i=1}^{N} (y_i - \bar{y})^2$; this is a measure of the total variation among all the observed y-values, and if it were divided by $(N - 1)$ would be like a variance: it is not quite true to say it would actually be a variance, for we shall see that one part of it can be used to estimate the variance of y while the other part tells us something else.

So far we have thought of the fitted line as $y = a + bx$; it is now clear that each original observed value $y_{i.\,Obs}$ is not *exactly* equal to $a + bx_i$, but to this plus a small term representing the deviation of the observed point from the fitted line (as in Fig. 11.1, where QP is a typical one of these deviations). Hence we write $y_i = a + bx_i + e_i$, where y_i and x_i stand for a pair of observations giving a point on the graph, and e_i is the deviation term. The assumption (2) thus requires that the e_i shall be mutually independent, not affected by x and shall all be taken from a $\mathcal{N}(0, \sigma^2)$ distribution.

The Analysis of Variance is laid out as in Table 11.1: the first column lists the components into which the total sum of squares is split. These are: one part measuring how much of the variation among the y_i can be explained by the fitted regression line (i.e. the extent to which this variation was merely a linear relation with x), and a second part measuring the size of deviations from this line. It is this second part that is really of interest, for if it is small the line will be a good fit. However, the calculation is most easily made by working out S and the sum-of-squares for regression, S_R; and then the difference between these, $S_D (= S - S_R)$, is the required measure of deviation. (It is, in fact, equal to $\sum_{i=1}^{N} (y_{i.\mathrm{Obs}} - y_{i.\mathrm{Pred}})^2$, or $\sum_{i=1}^{N} e_i^2$, though it is very rarely calculated in this way.) S has already been defined: it is calculated as $S = (1/N)[N \sum_{i=1}^{N} y_i^2 - G_y^2]$.

S_R is found to be $\{\sum_{i=1}^{N}(x_i - \bar{x})(y_i - \bar{y})\}^2 / \{\sum_{i=1}^{N}(x_i - \bar{x})^2\}$, and is calculated as

$$S_R = \frac{[(1/N)(N \sum_{i=1}^{N} x_i y_i - G_x G_y)]^2}{(1/N)(N \sum_{i=1}^{N} x_i^2 - G_x^2)}$$

For Example 11.1, the results are:

$$S = \tfrac{1}{8}(8 \times 9536 - 234^2) = 21532/8 = 2691 \cdot 5$$

$$S_R = \frac{[\tfrac{1}{8}(8 \times 2348 - 57 \times 234)]^2}{\tfrac{1}{8}(8 \times 579 - 57^2)} = \frac{680 \cdot 75^2}{172 \cdot 875} = 2680 \cdot 6685$$

Then $S_D = S - S_R = 10 \cdot 8315$.

In Chapter 8, we found that the sum of squares of N unit normal deviates was distributed as χ^2, and that its degrees of freedom were $(N-1)$; further, that if the normal deviates have variance σ^2 rather than unity, the numerical value of this χ^2 divided by its degrees of freedom, $(N-1)$, provides an estimate of σ^2. In the present case, the sum of squares for deviations from regression, S_D, is equal to $\sum_{i=1}^{N} e_i^2$; we have assumed that e_i is $\mathcal{N}(0, \sigma^2)$, and therefore S_D is distributed as χ^2. A suitable Null Hypothesis for this problem states that the regression coefficient b is zero. If that is so, S and S_R are also distributed as χ^2; S, being based on N observations, is $\chi^2_{(N-1)}$, but it is not at once clear what will be the d.f. for S_R and S_D. A theorem in the mathematical development of the method of least-squares may be applied to discover that S_D is $\chi^2_{(N-2)}$, and a further theorem, this time about the sums of χ^2 variates, tells us that S_R is therefore $\chi^2_{(1)}$. A good working rule, which applies to more complicated regression equations also, is that S_R has d.f. equal to one less than the number of constants needing to be estimated: in this Example we have estimated a and b. The d.f. of all the components in an Analysis of Variance table (here S_R and S_D) must add to the d.f. of S; hence S_D must have $(N-2)$ d.f.

The second column of Table 11.1 contains the d.f., and the third column the sums of squares S, S_R and S_D. On the N.H., each of the χ^2-variates just mentioned will, when divided by its d.f., give an estimate of σ^2: so the fourth column contains the **mean squares** $S_R/1$, $S_D/(N-2)$ (mean square being the general name for a sum of squares divided by its d.f.). The ratio of two such estimates of σ^2 is distributed as F; specifically, $(S_R/1)/(S_D/(N-2))$ is $F_{(1, N-2)}$. If this F is significantly large, we reject the N.H., and assume that b is not zero. It can be shown that if b is *not* zero, the value of the regression mean square must be greater than σ^2, so the correct F-test is the one-tailed form, as tabulated in Table III. Whether or not b is zero, the deviations-mean-square $S_D/(N-2)$ provides a valid estimate of σ^2, the variance of e_i.

Table 11.1   Algebraic form of Analysis of Variance for testing Regression Goodness-of-Fit

Source of variation	D.F.	S.S.	M.S.	Test
Regression	1	S_R	$M_R = \dfrac{S_R}{1}$	$\dfrac{M_R}{M_D} = F_{(1, N-2)}$
Deviations from line	$N-2$	$S_D = S - S_R$	$M_D = \dfrac{S_D}{N-2}$	
Total	$N-1$	S		

Table 11.2 shows this method of analysis applied to the data of Example 11.1. The ratio of the two mean squares is very large: reference to the F-table with 1 and 6 d.f. shows the actual value to be very much greater than that needed for significance at 0.1%, and so we must reject the N.H. We may claim that b is significantly different from 0.

In Table 11.2 and later, we use the shorthand that significance at the 5%, 1% and 0.1% levels is indicated by one, two and three asterisks respectively, and n.s. denotes not significant.

Table 11.2   Analysis of Variance for Example 11.1

Source of variation	D.F.	S.S.	M.S.	
Regression	1	2680·6685	2680·6685	$F_{(1, 6)} = 1484 \cdot 9$***
Deviations	6	10·8315	1·8053	
Total	7	2691·5000		

The most interesting feature about a regression calculation will usually be the numerical value of the slope b, and so it is useful to be able to set limits to b. The necessary estimate of σ^2, as we have just seen, is obtained from the deviations-mean-square. Further, the distribution of $\hat{b}$ can be shown to be normal, and so confidence limits can be calculated in the manner described at the beginning of Chapter 9 (p. 77) for normal variates. The variance of $\hat{b}$ is estimated by $\hat{\sigma}^2/[\sum_{i=1}^{N}(x_i-\bar{x})^2]$, where we use the symbol $\hat{\sigma}^2$ to denote that estimate of σ^2 provided by the deviations-mean-square. This estimate $\hat{\sigma}^2$ has $(N-2)$ d.f., and so formula (**A**) of Chapter 9 may be applied to give the 95% limits for b as

$$\hat{b}-t_{(N-2,\,0.05)}\sqrt{\frac{\hat{\sigma}^2}{\sum_{i=1}^{N}(x_i-\bar{x})^2}} \leqslant b \leqslant \hat{b}+t_{(N-2,\,0.05)}\sqrt{\frac{\hat{\sigma}^2}{\sum_{i=1}^{N}(x_i-\bar{x})^2}} \quad \textbf{(E)}$$

In Example 11.1, $\hat{b}=3.94$, $\hat{\sigma}^2=1.8053$, and $\sum_{i=1}^{N}(x_i-\bar{x})^2=172.875$. Thus the variance of b is $1.8053/172.875 = 0.01044$, and its square root is 0.102. The d.f. for $\hat{\sigma}^2$ are 6; the values of $t_{(6)}$ at 5% and 1% are respectively 2.447 and 3.707. Therefore 95% limits for b are $3.94 \pm 2.447 \times 0.102$, i.e. 3.94 ± 0.25; so with 95% confidence we may say that the true value of the slope of the regression line relating y to x is not less than 3.69 nor greater than 4.19. To obtain 99% limits, 2.447 is replaced by 3.707 in the above calculation. Confidence limits to calculated values $y_i._{\text{Pred}}$ are not so easy to give, and a more advanced text should be consulted if these are needed.

The F-test above is really testing whether the regression-mean-square is considerably larger than the deviations-mean-square. We use the deviations-mean-square as the estimate of σ^2, but strictly speaking it is not the best estimate that could be found. If it is possible to read duplicate values of y at each value of x, we can obtain an estimate of σ^2 that genuinely measures only the random variation in y. In practice, this is not usually easy, or even possible, to arrange, and then we must rely on the deviations from linearity for a measure of σ^2. However, this ignores the possibility of there being a slightly systematic trend in the deviations, so small that the regression-mean-square is still much greater than the actual size of the deviations, but none the less apparent from a graph like Fig. 11.2. Here a straight line is, statistically, a very good fit; but the deviations from it are *systematic* and not random. No statistical test can make up for the failure to examine the data carefully before fitting a line. The best we can do is to test whether a line *fits* well, not whether it provides a reasonable hypothesis. The reason we take as Null Hypothesis for a significance test the value $b=0$ is that if there is no relation at all between y and x, then y is simply varying randomly about its mean $\bar{y}$, and is neither increasing (b positive) nor decreasing (b negative) with changes in x.

When it is clear that linear regression is not a satisfactory explanation of the relation between x and y (either from inspection of the graph or because $F_{(1, N-2)}$ is not significant), another useful approach for the biologist is to look for a *logarithmic* relation. If a curve like Fig. 11.3(a) appears, then $\log y = a + bx$ may fit, or if the curve is like Fig. 11.3(b) then it is worth examining $y = a + b \log x$. Thus if the original data seem to follow either of these two curves, a straight line may be obtained by plotting (a) y on a logarithmic scale and x on an ordinary one, or (b) y on an ordinary scale and x on a logarithmic one. To fit a line, the calculations proceed as described above except, of course, that in case (a) each value of y_i is replaced by its logarithm while x_i remains in its original units; or in case (b), y_i is unchanged and x_i is replaced by

Fig. 11.2   Slightly curvilinear regression relation between y and x.

$\log x_i$. There is nothing unnatural in using logarithmic units when necessary: indeed, Fig. 11.3(a) is often relevant where y is the size of a cell or culture and x is the amount of an added growth-stimulant substance present in the medium in which the cell or culture is growing. A similar growth curve also occurs, with x in this case measuring time, if the *rate* of growth of a cell or organism at any instant of time is proportional to its actual size at that instant. These curves are usually called *exponential*; the relation $\log_e y = a + bx$ is mathematically equivalent to $y = ce^{kx}$, in which c, k are two other constants replacing a, b and e^x is the exponential function that was employed in Chapters 5 and 6. (See Causton[1] for further explanation.) An example of Fig. 11.3(b) arises in insecticide and fungicide work, wherein the logarithm of the concentration, x, of applied insecticide or fungicide is often found to be linearly related to the numbers of insects or spores killed, y.

It is sometimes advocated that when $y = a + bx$ clearly does not fit, and the relation is obviously a curved one, a polynomial in x should be developed, $y = a + bx + cx^2 + dx^3 + \cdots$, higher powers of x continuing to be added until a sufficiently close fit is found. Apart from the arithmetical difficulty of fitting, it is usually an insuperable problem to explain the result at all sensibly; and where regression is being used to predict the behaviour of y in relation to x, on the basis of a number of observations, N, which may be quite small, this is very dangerous. It is even more dangerous to predict what might happen outside the range of the observations actually taken: for suppose that in Fig. 11.3(a) we had lacked the two points on the graph corresponding to the two lowest x-values. Without them, no real curvilinear tendency would be apparent, and so, no doubt, a straight line would be fitted. This would cross the x-axis to the right of o, and we might therefore waste much time searching for a reason why y should be zero while x is positive; but in fact this value of x lies outside the range of x-values available when the line was being determined.

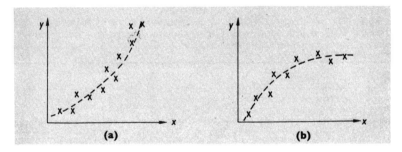

Fig. 11.3 (a) Relation between x, y which leads to the regression line $\log y = a + bx$; (b) relation leading to $y = a + b \log x$.

Sometimes it is not clear at once which of x, y is dependent on the other: a graph may look more like Fig. 10.1(a) or (d) than Fig. 11.1. A classical example is to let one variate represent the height of a father, and the other variate the height of his son. Can any meaning be attached to 'the regression of sons' heights on fathers' heights' or 'the regression of fathers' heights on sons' heights', and if so which is the more fundamental? Obviously sons' heights depend on fathers', and not vice versa, so sons' heights would be taken as y; but x cannot now be thought of as *fixed* in the same sense as in Example 11.1. So long as x, the fathers' heights, can be measured with errors which are negligible in relation to the heights being considered, this will not matter, and $y = a + bx$ can be fitted as above. But could there be any purpose in trying to fit $x = \alpha + \beta y$? As a method of predicting x_i, a father's height (presumed unknown), from y_i, the height of his son (presumed known), it might be useful. If, on these grounds, the calculation is carried out, the estimates $\hat{\alpha}$ and $\hat{\beta}$ will be those which make deviations from the line in the *horizontal* direction as small as possible, and the line $x = \hat{\alpha} + \hat{\beta}y$ will thus be different, perhaps considerably different, from the line $y = \hat{a} + \hat{b}x$, because β will have $\sum_{i=1}^{N}(y_i - \bar{y})^2$ in its denominator instead of the $\sum_{i=1}^{N}(x_i - \bar{x})^2$ appearing in $\hat{b}$. This emphasizes the importance of getting the logic of dependence and independence right before making calculations. If there seems to be no convincing reason for taking either variate as the independent one, it is perhaps unwise to try fitting so specific a relation as linear regression, but to rely on a correlation coefficient as a measure of relationship only, without introducing the concept of dependence at all.

EXERCISES

(For Answers and Comments, see p. 172)

11.1   Six fertilizer treatments were applied to plots of sugar beet, and the crop yield recorded for each. The treatments differed only in the amount of fertilizer applied, not in its constitution. Calculate the regression line

Treatment	(1)	(2)	(3)	(4)	(5)	(6)
Amount (cwt/acre) x	$\frac{1}{2}$	1	2	3	4	6
Yield (lb/plot) y	10	16	26	35	50	72

of y on x, test its goodness-of-fit, and find the predicted yields corresponding to each value of x used in the experiment and also to $x = 5$ and $x = 0$.
Comment on the latter.
Obtain also 95% confidence limits for the slope b of the line.

11.2   The following data represent the size of an organism at equally spaced times 0 to 8. Plot these on a graph and consider how one might proceed to find a regression relation between size (y) and time (x).

$$x = \quad 0 \quad\quad 1 \quad\quad 2 \quad\quad 3 \quad\quad 4 \quad\quad 5 \quad\quad 6 \quad\quad 7 \quad\quad 8$$
$$y = 0{\cdot}75 \quad 1{\cdot}20 \quad 1{\cdot}75 \quad 2{\cdot}50 \quad 3{\cdot}45 \quad 4{\cdot}70 \quad 6{\cdot}20 \quad 8{\cdot}25 \quad 11{\cdot}5$$

Calculate the equation of a suitable linear regression function.

11.3   The weekly salaries of seven scientists of different ages in a department are as shown below. Calculate the linear regression equation of y on x. What would happen to the slope and intercept if (i) all the scientists had a £10-per-week increase, (ii) they all had a 9% increase of salary?

Age, x years	20	22	28	33	36	45	55
Salary, £ y	69·8	78·5	108·8	134·0	144·0	182·5	198·6

12

The Principles of Experimental Design: the Completely Randomized Design

In research, comparisons must often be made between several sets of data collected from basically similar populations, such as when groups of plants of the same type have been grown under conditions alike except that different fertilizers were used for each group. We saw in Chapter 8 that, on suitable assumptions, *two* such sets of data may be compared by a *t*-test; and this needs extension for more than two sets. There are also considerations (practical as well as theoretical) about how to collect the data used for making comparisons, and these two problems give rise to the science of **Experimental Design and Analysis**.

In a classical example, suppose that a field has been sown as uniformly as possible with a standard variety of wheat, and has been marked off into a number, N, of **plots** of constant size. Several different fertilizer treatments are being compared; the treatments are usually labelled by capital letters A, B, C, D,... At the end of the season, the yield y of crop from each of the N plots is recorded. Consider what sources of variation there might be among these N different values of y; as we saw in the Introduction, there will be natural experimental error variation from plot to plot even when no different treatments are applied. But to this *random* source of variation must be added a second *systematic* source, due to the different effects of the treatments: suppose, for example, that A is a better-balanced fertilizer than B, but not so good as C which supplies the nutrient elements to the plant in a more easily available form in the soil. In this case, all the yields on plots receiving treatment C would tend to be higher than those with A, which in turn would be higher than those with B; though within each treatment natural variation will also be apparent.

In this example, as in many others in nature, the experimental-error distribution is normal (Chapter 6). Thus the yield distribution within any of the treatments has the same variance, σ^2, but the mean yields will be different; σ^2 is a characteristic of the type of plant which we are using but μ will vary depending on the fertilizer treatment. If μ_i is the mean yield per plot of the plants receiving treatment i, the distribution of plot yields for this treatment will be $\mathcal{N}\ (\mu_i,\ \sigma^2)$. We must recognize, however, that μ_i is not a constant in any absolute sense, for when any of the treatments is repeated on another site or in a subsequent season, under soil and climatic conditions which are inevitably not quite the same, the mean yield under that treatment will alter—sometimes considerably. But what is true of one treatment, A, say, will equally be true of B and of all the rest: all will react to the environmental change to much the same extent, so that the *differences between* pairs of treatments will be much more nearly constant, and independent of the exact conditions of the experiment. The statistical analysis therefore concerns itself with comparing, rather than simply estimating, the mean yields under different treatments.

The methods about to be developed have applications to a vast number of situations in biology, industry, medicine and the physical sciences. In animal nutrition, various diets will be under test on animals bred under uniform and controlled conditions; in the textile industry, various different combinations of temperature and humidity during the weaving of a cloth will constitute the treatments being compared; in analytical chemistry, different operators doing the same analysis will form the treatments, which here measure personal differences—if these are small, the method, being reproducible, will be more useful than one which permits larger operator-to-operator variation; in metallurgy, specimen components made from batches of alloy produced by different suppliers will be compared for strength, with a view to seeing if all suppliers produce equally good material. Treatments, in the general sense, just imply a source of systematic variation.

Any experiment to compare several treatments must embody at least the first two **Principles of Experimental Design**.

1 Since there is variation from plot to plot, the results from several unit plots must be used in assessing the response to each treatment: this is **Replication** of treatments, and besides giving satisfactory estimates of μ_i it enables σ^2 to be estimated from the variation among plots treated alike.

2 The treatment means will be genuinely unbiased only if no conscious allocation of better or poorer plots to the various treatments is made; this demands **Randomization**, i.e. random allocation of treatments so that each plot has the same chance of carrying any one of the treatments A, B, C, ... which are under test. Another way of saying this, in the language of Chapter 1, is that the selection of plots to carry each treatment is a random sample from all those plots available.

The simplest experimental design, which incorporates these two principles only, is the **Completely Randomized** layout. Suppose the treatments, called A, B, C, D, ..., are t in number, and that each one is replicated r times, the total number of experimental plots therefore being $rt = N$. Unless some treatments are of considerably more interest than others, it seems intuitively sensible to replicate each the same number of times, and it will be clear later that there are sound mathematical reasons for this. As a basis for the analysis of variance, set up the following model to explain how the yield (or whatever measurement has been taken per plot) arises:

$$y_{ij} = \mu + t_i + e_{ij} \tag{1}$$

In this, y_{ij} is the yield on the jth of the plots which carry treatment i (so that j takes values 1 to r), μ is the grand mean (average) yield over all the N plots, t_i is an *effect* due to treatment i (so that i takes values 1 to t), and e_{ij} is the natural variation or *experimental error* term on that particular plot. The effect t_i is a *deviation* from μ, positive for a better-than-average treatment, negative for a worse-than-average one; but since all these deviations are measured about the grand mean for the whole experiment they must themselves have mean 0, in other words their total $\sum_{i=1}^{t} t_i = 0$.

Whenever an Analysis of Variance is used, the error term e_{ij} must satisfy two conditions: (i) each e_{ij} is $\mathcal{N}(0, \sigma^2)$, (ii) all e_{ij}'s are mutually independent. Condition (i) implies that we must use material which, in the absence of different treatments, would give a normal distribution of yields, and also that the variance is the same for every treatment; this can reasonably be assumed unless some of the treatments have severe or unusual effects.

Condition (ii) is satisfied if the layout of treatments on plots really is random; if not, the yields and experimental errors on adjacent plots are likely to be correlated to some extent. Furthermore, the so-called *additivity* condition must hold: the mathematical model above must really consist of parts added together, not multiplied or combined in any other way. In Chapter 16 we consider how to proceed if some of these conditions do not hold.

THE ANALYSIS OF VARIANCE

This technique has two purposes: to find a valid estimate of σ^2 and to compare the mean yields under treatments A, B, C, D, ... A Total Sum of Squares S is defined, in the same way as in Chapter 11, to be the sum of all the squared deviations of individual plot yields from their overall mean, so that

$$S = \sum_{i=1}^{t} \sum_{j=1}^{r} (y_{ij} - \bar{y})^2$$

The symbol $\sum_{i=1}^{t} \sum_{j=1}^{r}$ implies that we add up all the squares obtained by letting both suffices i and j run through all their possible values, from 1 to t and 1 to r respectively; one $\sum$ is written for each suffix. In practical terms this means that we take the sum of squares over all the N plots in the experiment. This total sum of squares is now split into two parts, one corresponding to each of the terms in Model (1), namely treatments and error; note that μ has already been accounted for by measuring S about the grand mean $\bar{y}$.

As usual, let G = grand total of all experimental observations, so that in this case

$$G = \sum_{i=1}^{t} \sum_{j=1}^{r} y_{ij}$$

that is the sum of yields over all the N (=rt) plots. The grand mean yield in the whole experiment is $\bar{y} = G/N$ (or G/rt). If we use the symbol T_i to stand for the total yield of all the r plots which carried the treatment i, then

$$T_i = \sum_{j=1}^{r} y_{ij}$$

and the mean yield of all plots receiving treatment i is T_i/r. This is what we referred to as μ_i on page 102, but in Model (1) on page 103 we found it more useful to call it $\mu + t_i$. A *Sum of Squares for Treatments* is defined as

$$S_T = \sum_{i=1}^{t} \sum_{j=1}^{r} \left(\frac{T_i}{r} - \bar{y}\right)^2$$

using the deviations of the treatment means from the grand mean. When all treatment means are of similar size, and so all near in value to $\bar{y}$, then S_T will be small, but if S_T is large this implies that some means will be much larger than others and the whole set will need further examination. The *Error Sum of Squares*, S_E, is the difference between S and S_T: $S_E = S - S_T$; we shall always compute S_E by taking this difference (or an equivalent one in other designs, as in Chapter 13) and never by developing a special formula for it. However, the formulae for S, S_T are easy to express in terms of the y_{ij} and T_i,

$$S = \sum_{i=1}^{t} \sum_{j=1}^{r} (y_{ij})^2 - \frac{G^2}{N}$$

$$S_T = \frac{1}{r} \sum_{i=1}^{t} (T_i)^2 - \frac{G^2}{N}$$

In the problems we usually meet, where N is not very large, these formulae are simple and accurate to use; but they should not be used when programming for electronic computers.

EXAMPLE 12.1

Three fertilizer treatments, A, B, C, each applied to seven plots of strawberry plants, resulted in the following weights of crop (lb/plot):

A: 24, 18, 18, 29, 22, 17, 15;   total 143
B: 46, 39, 37, 50, 44, 45, 30;   total 291
C: 32, 30, 26, 41, 36, 28, 27;   total 220

The treatment totals ($T_A = 143$, $T_B = 291$, $T_C = 220$) add to give the grand total of all crop weights in the experiment, namely $G = 654$. There were three treatments each replicated seven times, i.e. $t = 3$ and $r = 7$; so $N = rt = 21$. Thus

$$S = (24^2 + 18^2 + 18^2 + 29^2 + 22^2 + 17^2 + 15^2 + 46^2 + 39^2 + 37^2 + 50^2$$
$$+ 44^2 + 45^2 + 30^2 + 32^2 + 30^2 + 26^2 + 41^2 + 36^2 + 28^2 + 27^2)$$
$$- \frac{654^2}{21}$$

$$= 22520 - \frac{427716}{21} = 22520 - 20367 \cdot 43 = 2152 \cdot 57$$

Also

$$S_T = \frac{1}{7}(143^2 + 291^2 + 220^2) - \frac{(654)^2}{21} = \frac{153530}{7} - 20367 \cdot 43$$

$$= 21932 \cdot 86 - 20367 \cdot 43 = 1565 \cdot 43$$

Hence $S_E = S - S_T = 587 \cdot 14$. Table 12.1 sets out the Analysis of Variance in symbolic form, and Table 12.2 the calculations for Example 12.1. Asterisks denote the level of significance, as on p. 96.

Table 12.1 Analysis of Variance for a Completely Randomized design

Source of variation	D.F.	Sum of squares	Mean square	
Treatments	$t-1$	S_T	$S_T/(t-1)$	$\dfrac{\text{Trt. M.S.}}{\text{Error M.S.}} = F_{[(t-1),\ t(r-1)]}$
Error	$t(r-1)$	S_E	$S_E/[t(r-1)]$	
Total	$rt-1$	S		

Table 12.2 Analysis of Variance for the data of Example 12.1

Source of variation	D.F.	Sum of squares	Mean square	F-ratio
Treatments	2	1565·43	782·72	$F_{(2,\ 18)} = 24\cdot00$***
Error	18	587·14	32·62	
Total	20	2152·57		

The value of σ^2 is estimated to be 32·62. The Null Hypothesis, of no effect of treatments, is rejected at the 0·1 % significance level.

Before we can proceed further with the Analysis of Variance, we must consider exactly what hypotheses we are testing statistically. Model (1) has already been set up (page 103); after having studied Chapters 7 and 8 readers will realize that this Model is in a form suitable to serve as an Alternative Hypothesis, which we shall accept only when we can show that some simpler Null Hypothesis is unreasonable. There is only one simpler idea that we can set up that is relevant to the present analysis, namely that the t_i can be omitted. So the Null Hypothesis will be

$$y_{ij} = \mu + e_{ij} \qquad (2)$$

This contains the random term only, and states exactly what we have assumed to be true if no effectively different treatments had been applied, namely that every y_{ij} would be $\mathcal{N}(\mu, \sigma^2)$. We now consider what can be said about S_T and S_E on this Null Hypothesis: they will both be sums of squares of normal deviates, so having χ^2 distributions (Chapter 8), with degrees of freedom $(t-1)$ and $t(r-1)$ respectively (the latter being equal to $(N-t)$, i.e. $rt-t$).

There is a simple working rule for obtaining degrees of freedom:

there are t treatments, and so it is possible to compare 1 with 2 (by calculating $t_2 - t_1$), 1 with 3, 1 with 4 and so on up to 1 with t, all of these comparisons, $(t - 1)$ in number, being independent of one another. But when we try to compare 2 with 3, this gives no fresh or independent information since $(t_3 - t_2)$ is just $(t_3 - t_1) - (t_2 - t_1)$ and these two comparisons have already been made; this is equally true for any other comparison attempted after the first $(t - 1)$. So the degrees of freedom, the number of independent comparisons possible among the treatments, equal $(t - 1)$, *one less than the actual number* of treatments. Similarly, among all the rt observations y_{ij}, $(rt - 1)$ independent comparisons can be made, so that S has d.f. $(rt - 1)$. As S_E is the difference between S and S_T, so its d.f. are (d.f. of S) *minus* (d.f. of S_T); error-degrees-of-freedom in more complicated types of experimental design are always calculated by taking away from the d.f. of S all the d.f. of the other sums of squares contained in S (see Chapter 13).

As we saw in Chapter 11, the *mean squares* for each term in the Analysis of Variance are useful: these are the sums of squares divided by their corresponding degrees of freedom. Since S_T is $\chi^2_{(t-1)}$, on the Null Hypothesis, $S_T/(t-1)$ is an estimate of σ^2; similarly $S_E/[t(r-1)]$ also estimates σ^2. Therefore, on the Null Hypothesis the ratio *treatments-mean-square/error-mean-square* is distributed as $F_{[(t-1), t(r-1)]}$ since it is the ratio of two expressions each of which estimates σ^2. Hence (Chapter 8) the numerical value of this ratio should not differ significantly from 1, and if it does so we shall reject the Null Hypothesis upon which this theory has been built. That is to say, the simple model $y_{ij} = \mu + e_{ij}$ will not be satisfactory. Having rejected the Null Hypothesis, we accept the Alternative Hypothesis, Model (1). This process is automatic, for we never actually test anything about the Alternative. But, in fact, it is a reasonable one for situations where the F-ratio *does* exceed 1, because we can show that on the Alternative Hypothesis the treatments-mean-square estimates $(\sigma^2 + r\sum_{i=1}^{t} t_i^2/(t-1))$, and this must exceed σ^2 since the t_i are not all zero (as they are on the Null Hypothesis). The error-mean-square still estimates σ^2; it therefore gives us the necessary estimate of σ^2 whichever hypothesis we decide is appropriate.

At the outset, we aimed to estimate σ^2, which we have now done, and also to compare the various treatments. So far, we have seen only whether the treatments can be considered a homogeneous set or not: if not (i.e. when F is significant), we want to look at the so-called *treatment means*—the mean yields under the various treatments—and especially, as we saw earlier, at their differences. For *any pair* of means, a test of the significance of their difference is provided by the *t*-test (Chapter 8, p. 61), since each mean has been formed from r observations from a normal distribution of variance σ^2. So for A and B

$$\frac{T_A/r - T_B/r}{\sqrt{2\sigma^2/r}}$$

is distributed as $t_{(f)}$, where f denotes the error d.f.; f is equal to $t(r-1)$ in the completely randomized design. Writing $t_{(0.05,\ f)}$ to stand for the value shown in the t-table (Table I) at the 5% significance level and with f degrees of freedom, we see at once that if

$$\frac{T_A/r - T_B/r}{\sqrt{2\sigma^2/r}}$$

exceeds $t_{(0.05,\ f)}$, the two means differ significantly at 5%. This is exactly the same as to say that there is significance when $T_A/r - T_B/r$ exceeds $(t_{(0.05,\ f)} \times \sqrt{2\sigma^2/r})$, the latter expression being called the **significant difference** (sometimes *least significant difference*) between the two means. When all the treatments have, as recommended, been replicated equally (r times) the same calculation is required for testing significance between *any* two means; therefore the final step in an analysis wherein F has proved significant is to work out a significant difference (usually at each of the significance levels 5%, 1% and 0·1%) and use it to compare the treatment means. Any pair of means whose difference is greater than the significant difference may be declared significantly different.

EXAMPLE 12.1 (concluded)

The treatment means are, for A, $143/7 = 20·43$, for B, $291/7 = 41·57$, for C, $220/7 = 31·43$. The error-mean-square estimate of σ^2 is $\hat{\sigma}^2 = 32·62$. The error d.f., $f = 18$, and $t_{(0.05,\ 18)} = 2·101$. The expression $\sqrt{2\hat{\sigma}^2/r}$ is $\sqrt{2 \times 32·62/7} = \sqrt{9·32} = 3·05$, giving a significant difference of $2·101 \times 3·05 = 6·41$ at the 5% level. Similarly $t_{(0.01,\ 18)} = 2·878$, leading to a significant difference of 8·78, and $t_{(0.001,\ 18)} = 3·922$, leading to a significant difference of 11·96. Now the means of A, C differ by 11·00, which is greater than 8·78 but not so great as 11·96; so A, C differ significantly at the 1% level but not at 0·1%. Likewise, A, B differ at 0·1%, and B, C at the 1% level.

The reader may have spotted an anomaly in the previous paragraph, since there are only 2 d.f. for treatments and yet three comparisons have been made. There is, in fact, a greater fundamental difficulty even than this, for if we do make comparisons between all possible pairs of means we violate the level of significance (5% etc.) that we claim to be using, in the sense that our results are significant not at that level but at some much less worth-while one. The actual size of level depends on how many treatments are included in the experiment and which pair we test. The unthinking practice of testing the largest treatment against the smallest can give some wildly wrong answers and, indeed, if the experiment contains enough treatments—even though these may form a homogeneous set—a t-test between the largest and smallest is highly likely to show a significant difference. So as to reduce the risk of making wrong statements after carrying out the analysis, various safeguards have been suggested.

Firstly, note that in Example 12.1 an F-test was carried out first, and only when F proved significant did we proceed further, to t-tests or significant differences. Provided this order is always observed, serious errors are not likely. But, even then, we cannot properly compare every pair of means, and certainly it is unwise to make a larger number of statements about treatment differences than there are degrees of freedom for treatments (for not all of these statements can be independent of each other, to give fresh information). And such comparisons as are tested should have been decided on *before* the experimental results are seen rather than be suggested by them: in other words, we had some definite ideas that we wished to examine when embarking on the experiment—this is certainly not asking too much of a properly planned piece of research. Often there will be a relatively small number of comparisons that are of use or interest, for example: (1) one treatment may be a *control*, either untreated or having a well-known, established treatment applied to it (such as a standard fertilizer) and the only interest is to compare the other treatments with this one; (2) the treatments can be split up into groups, such as when a plant-breeder puts a large number of new crosses, raised at the same time, into the same experiment—the treatments here are the different crosses—but really wants to compare only crosses which have a parent in common; (3) the treatments form a *factorial* set, considered in Chapter 15. When only a small number of comparisons have been nominated for testing, it is possible to omit F-tests and proceed to t-tests at once from the estimate of σ^2 in the Analysis of Variance. In this situation, the sole reason for putting a mixed-looking set of treatments into the same experiment may be that it is convenient to have only one piece of work in progress on the same type of plant at the same time, and to obtain the estimate $\hat{\sigma}^2$ from as many plots as possible. If some more specific hypothesis about the relation of y to the treatments is possible, such as a regression (Chapter 11) or other form of response curve (Chapter 15, pp. 134–5), this will of course be tested in preference to making large numbers of individual treatment comparisons.

Secondly, significances or significant differences are sometimes deliberately not quoted in published results of experiments, preference being given (see Chapter 9) to the *standard error of a treatment mean*,

$\sqrt{\hat{\sigma}^2/r}$, or to the *standard error of a difference between two means*, $\sqrt{2\hat{\sigma}^2/r}$. Of course, in this as in all other methods, an Analysis of Variance table is prepared, because it is the best way of obtaining the necessary estimate of σ^2. But since comparisons will inevitably be made by people wishing to apply the results of these experiments, significant differences will, in practice, be calculated in any case, by multiplying the first of these standard errors by $(t_{(0.05, f)}\sqrt{2})$ or the second by $t_{(0.05, f)}$. However the results are presented, the temptation to extract too many comparisons still has to be avoided. And of course the remarks of Chapter 7 about the significance level used (5%, 1%, etc.) apply equally here.

Thirdly, there may be occasions when it really is useful to compare the whole set of treatments. Usually this will only be the right approach in the early stages of a programme of work, with a view to choosing what treatments should be included in follow-up studies—where some treatments will eventually be discarded. Statistical evidence of which seem the best and worst treatments can be a useful addition to other biological considerations in deciding what should be included in the next stage of the programme: in the plant-breeding example mentioned earlier, those high-yielding crosses which have good growth habit and are disease resistant will be examined further. The plant breeder will not look *only* at the statistics of yield, of course; he will use all his biological knowledge, and observe as well as measure properties that seem to him desirable. Various statistical tests have been proposed for carrying out indiscriminate comparisons of all treatments; *t*-tests cannot be used for the reasons given above. One of the more commonly used is Duncan's test (called a 'multiple-range' test). First σ^2 is estimated in the usual way, then a list of the means is written down in increasing order of size, and the final aim is to form groups of means that do not differ from one another significantly. To do this, special tables are needed; these, and a detailed description of the method, may be found in Duncan's original paper[4] or in the book by Edwards.[5] The practice, sometimes adopted in journals, of using Duncan's test instead of *t*-tests when reporting the final results of a programme of planned experiments, is not correct, especially when the treatments form a carefully structured set such as a factorial (Chapter 15). Duncan's test, **and others like it, apply at the stage in the programme where possible** treatments are being explored and 'screened', not at the time when specific hypotheses have been set up for testing.

LAYOUT OF THE EXPERIMENT

To satisfy the principle of randomization, we number the available field plots 1 to 21 and arrange that a random selection of seven of the numbers between 1 and 21 carries A, a further random seven, B, and the rest, C. In order to use the run of random digits quoted in Chapter 1, mark these off in pairs: 07, 43, 55, 27, 18, 34, 94, 56, ... Now 07 corresponds to plot number 7, so we write this down first in our list; then for numbers greater than 21, let 22–42 correspond respectively to 1–21, likewise 43–63 and 64–84 also correspond to 1–21, while 85 upwards (and 00) are ignored. Thus 43 corresponds to 1; 55 to 13; 27 to 6; 18 is itself; 34 corresponds to 13, but this has already appeared once; 94 is ignored; 56 corresponds to 14, and so on. Neglecting all repeats, we arrive by this process at a list of numbers, between 1 and 21, in random order: 7,1,13,6,18,14,2; 19,3,11,21,16,17,5; 8,20, ... This may be called a **random permutation** of the numbers 1–21, and such permutations are printed by Fisher and Yates[7] and Cochran and Cox[3], though not for numbers above 20. The first seven plots in the list are allocated to A, the next seven to B, and the rest to C, as in Fig. 12.1.

This layout may look rather systematic, to the extent that the treatments are somewhat bunched, especially A and C: had we been producing a layout 'haphazardly, out of the head', we would hardly have accepted this one. But it is a perfectly valid one, satisfying both principles so far laid down. If there are valid reasons for suspecting syste-

1 A	2 A	3 B	4 C	5 B	6 A	7 A
8 C	9 C	10 C	11 B	12 C	13 A	14 A
15 C	16 B	17 B	18 A	19 B	20 C	21 B

Fig. 12.1 Completely randomized layout for seven replicates of three treatments.

matic soil variation, so that we prefer a more even scatter of treatments through the field, the third Principle, that of **Blocking**, is used; wherever possible the layout will also satisfy the fourth Principle, **Orthogonality**. These two further principles are set out in the next chapter.

ASSUMPTIONS IN THE ANALYSIS OF VARIANCE

There are three basic assumptions which must hold, at least to a very good degree of approximation, if the Analysis of Variance technique is to be valid. It is used so often that perhaps these are taken too much for granted.

First (see page 103), the Model set up to explain the observations—Model (1) and others like it in later chapters—must be *additive*. Second, the measurement which we analyse must be *normally* distributed, or very nearly so; a skew distribution for y would make the results very unreliable. Third, σ^2 must be *constant* for all plots, that is to say, also, for all treatments. It is not easy to test for the first two of these conditions, though in Chapter 16 (page 146) we consider practical ways of analysing non-normal data. We also consider non-constant variance in Chapter 16, but for this there *is* a test available.

Bartlett's test for homogeneity of variance takes k sets of data, from each of which an estimate of variance has been calculated; the ith estimate is s_i^2, on f_i degrees of freedom. First compute an 'average' variance

$$s^{*2} = \sum_{i=1}^{k} f_i s_i^2 \bigg/ \sum_{i=1}^{k} f_i. \quad \text{Then obtain} \quad M = \left(\sum_{i=1}^{k} f_i\right) \log_e s^{*2} - \sum_{i=1}^{k} (f_i \log_e s_i^2) \quad \text{and}$$

$$C = 1 + \frac{1}{3(k-1)} \left(\sum_{i=1}^{k} \left(\frac{1}{f_i}\right) - 1 \bigg/ \sum_{i=1}^{k} f_i\right). \quad \text{Finally } M/C \text{ is tested as } \chi^2_{(k-1)}.$$

EXAMPLE 12.2

In four sets of data, estimated variances are 14·5 (on 8 d.f.), 35·4 (6 d.f.), 67·2 (23 d.f.) and 23·3 (12 d.f.). Do these differ significantly?

$$s^{*2} = (8 \times 14\cdot5) + (6 \times 35\cdot4) + (23 \times 67\cdot2) + (12 \times 23\cdot3)/(8 + 6 + 23 + 12)$$
$$= 2153\cdot6/49 = 43\cdot9510$$
$$M = (49 \times 3\cdot7831) - \{(8 \times 2\cdot6741) + (6 \times 3\cdot5667) + (23 \times 4\cdot2077)$$
$$+ (12 \times 3\cdot1485)\} = 8\cdot0193.$$

$C = 1 + \frac{1}{9} \left(\frac{1}{8} + \frac{1}{6} + \frac{1}{23} + \frac{1}{12} - \frac{1}{49}\right) = 1\cdot0442$, so $M/C = 7\cdot68$ which is not quite significant when tested as $\chi^2_{(3)}$, so that we cannot reject the Null Hypothesis that all variances are the same. Unfortunately, variances do have to be very different before the test shows significance, and non-normality of data has a serious effect on the test. Exercises 16.4 and 16.5 use this method. (Note that if logs to base 10, rather than base e, are used, the expression given in the formula for M must be multiplied by 2·3026.)

EXERCISES

(For Answers and Comments, see p. 172)

12.1 A sample of plant material is thoroughly mixed, and 15 aliquots taken from it for determination of potassium content. Three laboratory methods A, B, C are employed, A being the one generally used. Five aliquots are analysed by each method, giving the following results (μg/ml).

> A: 1·83, 1·81, 1·84, 1·83, 1·79
> B: 1·85, 1·82, 1·88, 1·86, 1·84
> C: 1·80, 1·84, 1·80, 1·82, 1·79

Examine whether methods B and C give results comparable to those of method A.

12.2 Eight varieties, A–H, of blackcurrant cuttings are planted in square plots in a nursery, each plot containing the same number of cuttings. Four plots of each variety are planted, and the shoot length made in the first growing season measured.

The plot totals (m) are:

> A: 46, 29, 39, 35   E: 16, 37, 24, 30
> B: 37, 31, 28, 44   F: 41, 28, 38, 29
> C: 38, 50, 32, 36   G: 56, 48, 44, 44
> D: 34, 19, 29, 41   H: 23, 31, 29, 37

B and C are standard varieties; assess the remaining 6 for vigour in comparison with B and C.

13

Randomized Blocks and some extensions

BLOCKING

If it is known or suspected that there is another source of systematic variation, in addition to and independent of the treatments, an extra term is included in the Model, giving

$$y_{ij} = \mu + t_i + b_j + e_{ij} \qquad (3)$$

In the example of a field experiment, this new source of variation will appear in the form of trends in fertility over the whole of the experimental area; if it is possible to find smaller areas each homogeneous within itself, these are used as **blocks**, and every block contains one complete set of the treatments, so that it must consist of t plots. Since there are to be r replicates of each treatment, there must be r blocks, and the suffix j in model (3) takes on a specific meaning: plot (ij) will now be that which carries treatment i in block j, and b_j is an *effect* due to blocks, having a similar character to t_i which measures treatments. The difference between blocks is made to account for as much as possible of the systematic soil variation, consistent with keeping block size at t plots. Fig. 13.1 indicates how five treatments, each replicated four times, would be laid out in blocks to remove the effect of a fertility trend running from top to bottom of the diagram; following the usual convention we label the treatments by capital letters, and blocks in roman numerals.

With the trend shown, of soil conditions improving from the top to

Fig. 13.1   Randomized block layout for four replicates of five treatments on land having a fertility trend in the direction shown.

the bottom of Fig. 13.1, all plots in block I are naturally slightly inferior to those in block II, which in turn are inferior to those in block III, etc., so that in the absence of different treatments we would expect the yields in I to be systematically lower than those in II, and so on. Note that unless blocks run at right angles to the trend, all plots in each particular block are not equally affected by the trend, and b_j does not have the meaning intended. And finally, if one of the blocks is very different from the rest, its effect may not be independent of treatments: one very poor block of soil may induce a worse reaction to the less good fertilizers (in the language of Chapter 15, an *interaction* between blocks and treatments), and this invalidates Model (3).

Within each block, plots must be as closely comparable as possible (blocks are to be homogeneous within themselves) and so the blocks of Fig. 13.1 are subdivided in such a way as to make each plot equally affected by the trend. If the field is patchy, not showing a steady trend like this, then the blocks should be of such a shape as to remove as much patchiness as possible. The layout of treatments must be *random* within each block, a fresh randomization being used for each block. For this purpose, tables of random permutations (referred to in Chapter 12) are often useful, but again it is possible to work directly from random numbers. Using the same run as in Chapters 1 and 12, we shall order the digits 1, 2, 3, 4, 5 at random in four different ways, and then apply these to the treatments A, B, C, D, E in blocks I, II, III, IV. Waste of digits can be avoided by making 6 correspond to 1, 7 to 2, 8 to 3, 9 to 4, 0 to 5, so the run 07435527183494562314 becomes 52435522133444451 2314. The first four digits here are 5243, so giving 52431 as a random permutation of 1 to 5; then 552213 gives 52134; 344451 gives 34512; 2314 gives 23145. Translating to letters, we have 52431 giving EBDCA, and the others are EBACD, CDEAB and BCADE.

Blocks, like treatments, really mean only *a source of systematic variation*, and the name, though derived from the original application to field trials, may imply different batches of material in an industrial experiment, different litters of animals in a comparison of diets, or different machines or operators in a programme of analyses in a chemical laboratory. Even when there is no other reason for introducing blocks into a layout, they may still be put in to provide convenient units representing a day's work, or a morning's work or some other suitable amount of recording time: this helps if one person is not able to complete the job or if weather intervenes and makes the recording of a field experiment spread over a few sessions or a few days. In these cases, blocks may help to maintain homogeneity, and certainly will not reduce it.

In the analysis, S and S_T are the same as for a completely randomized layout. We need to define $B_j =$ the total of all yields of plots in block j, so that the mean yield in block j is B_j/t (the block contains t plots). The total number of plots is $N = rt$. The Blocks Sum of Squares is

$$S_B = \frac{1}{t} \sum_{j=1}^{r} (B_j)^2 - \frac{G^2}{N}, \quad \text{or} \quad \frac{1}{N}\left(r \sum_{j=1}^{r} B_j^2 - G^2\right)$$

and will have $(r-1)$ degrees of freedom by the same argument as previously used. All the description in Chapter 12 of methods of testing for treatment differences can be carried through for block differences though, in the field, comparisons between block means are unlikely to be wanted: it will suffice to know that the blocks have done their job of removing some systematic variation. The error-sum-of-squares is $S_E = S - S_T - S_B$, and will have d.f. $(rt-1)-(t-1)-(r-1) = rt-r-t+1$, which simplifies to the product $(r-1)(t-1)$; as usual the error-mean-square supplies the necessary estimate of σ^2. Table 13.1 gives the Analysis of Variance for a Randomized Block design.

Table 13.1   The Analysis of Variance for a Randomized Block Design

Source of variation	D.F.	Sum of squares	Mean square	F-ratio
Blocks	$(r-1)$	S_B	$S_B/(r-1)$	$\dfrac{\text{Blocks M.S.}}{\text{Error M.S.}}$ is $F_{[(r-1),\ (r-1)(t-1)]}$
Treatments	$(t-1)$	S_T	$S_T/(t-1)$	$\dfrac{\text{Trts. M.S.}}{\text{Error M.S.}}$ is $F_{[(t-1),\ (r-1)(t-1)]}$
Error	$(r-1)(t-1)$	$S_E = S - S_B - S_T$	$S_E/(r-1)(t-1)$	
Total	$(rt-1)$	S		

EXAMPLE 13.1

Samples from five different suspensions of bacteria A, B, C, D, E, were examined under a microscope by four different observers, I, II, III, IV; the order in which each observer dealt with the samples was randomized to reduce errors due to fatigue, and the numbers of organisms recorded from the samples were as summarized in Table 13.2. (Note that if observer I worked in the order EBDCA, II worked EBACD, III worked CDEAB and IV worked BCADE, Fig. 13.1 would exactly represent the scheme followed—observers are *blocks*.)

Table 13.2   Data of Example 13.1

Observer number	Suspension					Observer's total
	A	B	C	D	E	
I	68	71	54	95	73	361
II	82	78	67	116	85	428
III	77	74	65	103	88	407
IV	59	70	54	90	76	349
	286	293	240	404	322	1545 = *Grand Total*

In this example, we may be interested in the actual comparison of observers, as well as in eliminating systematic differences due to them; and comparisons between certain of the suspensions will have been specified for testing. An Analysis of Variance is set up following the scheme of Table 13.1, and the details of this are given in Table 13.3.

Table 13.3   Analysis of data in Example 13.1

Source of variation	D.F.	S.S.	M.S.	F-tests
Blocks = Observers	3	839·75	279·917	$F_{(3, 12)} = 17·40$ ***
Treatments = Suspensions	4	3685·00	921·250	$F_{(4, 12)} = 57·28$ ***
Error	12	193·00	16·083	
Total	19	4717·75		

Estimate of $\sigma^2 = 16·083$; the effects of Blocks and Treatments are both significant at the 0·1% level.

$$S_T = \tfrac{1}{4}(286^2 + 293^2 + 240^2 + 404^2 + 322^2) - \tfrac{1}{20}(1545^2), \quad \text{reducing} \quad \text{to}$$
$$3685{\cdot}00;$$
$$S_B = \tfrac{1}{5}(361^2 + 428^2 + 407^2 + 349^2) - \tfrac{1}{20}(1545^2), \text{ reducing to } 839{\cdot}75;$$
$$S = (68^2 + 71^2 + \cdots + 90^2 + 76^2) - \tfrac{1}{20}(1545^2), \text{ the first term contain-}$$
ing the squares of all the individual counts; S reduces to $4717{\cdot}75$.

Since there are 4 observers, S_B has 3 d.f.; similarly S has 19 d.f. and S_T has 4 d.f. On subtracting, $S_E = 193{\cdot}00$ and has 12 d.f.

Significant differences at the 5% level between means for observers (blocks) are $t_{(0{\cdot}05,\ 12)}\sqrt{2\hat{\sigma}^2/5}$, because each observer-mean is based on five records and the estimate of σ^2 on 12 degrees of freedom; differences at the 1% and 0·1% levels may be calculated similarly. Means for suspensions are based on four records each, and so significant differences will be $t_{(0{\cdot}05,\ 12)}\sqrt{2\hat{\sigma}^2/4}$ etc. Now $\sqrt{2 \times 16{\cdot}083/5} = \sqrt{6{\cdot}433} = 2{\cdot}54$, and $\sqrt{2 \times 16{\cdot}083/4} = \sqrt{8{\cdot}042} = 2{\cdot}84$, while the values of $t_{(12)}$ from tables are 2·179 (5%), 3·055 (1%) and 4·318 (0·1%). The complete list of significant differences for observers is thus $2{\cdot}54 \times 2{\cdot}179 = 5{\cdot}53$ at 5%, $2{\cdot}54 \times 3{\cdot}055 = 7{\cdot}76$ at 1% and $2{\cdot}54 \times 4{\cdot}318 = 10{\cdot}97$ at 0·1%. Means for observers are: I, 72·2; II, 85·6; III, 81·4; IV, 69·8. Likewise, significant differences for suspensions are $2{\cdot}84 \times 2{\cdot}179 = 6{\cdot}19$ at 5%, $2{\cdot}84 \times 3{\cdot}055 = 8{\cdot}68$ at 1% and $2{\cdot}84 \times 4{\cdot}318 = 12{\cdot}26$ at 0·1%. Means for suspensions are: A, 71·5; B, 73·3; C, 60·0; D, 101·0; E, 80·5.

After the discussion of Chapter 12, the reader will be rather wary of making all possible comparisons between the observer means or the suspension means. As regards suspensions, D is very much the greatest, significantly higher at 0·1% than any other. Also A and B are not very different from one another, while C appears to be lower than these two. If these results, or such parts of them as have been covered by the hypotheses to be tested, are sensible, they will be accepted with no further study; but if anomalies appear, such as if C had been expected to be very like D, the conduct of the experiment needs examination. The result of such examination is sometimes to discover sources of serious error, e.g. the suspensions not being properly shaken, and sometimes to reconsider the hypotheses. The comparison between observers in this example suggests that I and IV are different from II and III: explanations of these types of result would be sought in terms of skill, care or experience.

Orthogonality

This is not an essential principle of experimentation, but a very desirable one; the proper definition involves mathematical ideas outside our scope, but the implications of orthogonality are not hard to grasp. In randomized blocks, every block contains every treatment just once, so that if one block happens to be different from the others, every treatment is similarly affected. Had this particular block not contained every treatment, some adjustment would have had to be made to those treatments that were missing from it; it would not have been possible to consider treatment means by ignoring blocks—nor, indeed, block means by ignoring treatments, since the treatments absent from a particular block might, in their turn, have differed appreciably from those present in it. In an *orthogonal* design, each classification (blocks and treatments, also rows and columns in Latin Squares, below) can be examined independently of every other one.

LATIN SQUARES

A randomized block removes one systematic source of variation additional to treatments; a Latin Square removes two such sources. In a field experiment, if two probable fertility trends can be thought of, in directions at right angles, both need to be made the basis of blocking. Thus when there is a slope in the land being used, and also a climatic trend (e.g. effects of wind, rain) at right angles to this, a randomized block cannot take out all the known variation. The Latin Square, previously studied by mathematicians purely for its pattern, has been applied to this situation. It is a layout in which every letter (A, B, C, ..., the set of treatments) occurs once in each row and once in each column, as illustrated in Fig. 13.2; this shows that the square must contain t rows and t columns, and hence t replicates of each treatment, and also $N = t^2$.

Fig. 13.2   Latin square layout for five treatments.

Analysis proceeds along the same lines as for randomized blocks, but instead of the one sum of squares for blocks, systematic variation is now taken out by two sums of squares which are always called Rows (S_R) and Columns (S_C). The mathematical Model will now read

$$y_{ijk} = \mu + t_i + r_j + c_k + e_{ijk} \tag{4}$$

in which t_i refers to treatments, as usual, and the two terms r_j, c_k to rows and columns respectively in the layout; e_{ijk} is an error-term with all the usual properties. The use of three suffices i, j, k, is sometimes a source of confusion, especially as each may take the values 1 to t; but a glance at Fig. 13.2 will make it clear that when i and j have been chosen (e.g. treatment A in row 1), k cannot in fact take any more than one value (column 3) and the correct number of plots, t^2, is thus accounted for. If we define R_j, C_k as the total of the yields of all plots in the jth row and kth column respectively in the layout, the sums of squares required are:

$$S_R = \frac{1}{t} \sum_{j=1}^{t} R_j^2 - \frac{G^2}{t^2}; \qquad S_C = \frac{1}{t} \sum_{k=1}^{t} C_k^2 - \frac{G^2}{t^2}$$

S and S_T are similar to their previous forms, and since $r = t$, these become:

$$S_T = \frac{1}{t} \sum_{i=1}^{t} T_i^2 - \frac{G^2}{t^2}; \qquad S = \sum_{i=1}^{t} \sum_{j=1}^{t} \sum_{k=1}^{t} y_{ijk}^2 - \frac{G^2}{t^2}$$

and finally S_E will be $S - S_T - S_R - S_C$. The degrees of freedom for S_T, S_R, S_C are each $(t-1)$; for S, (t^2-1), and so for S_E, $(t^2-1) - 3(t-1)$, reducing to $(t-1)(t-2)$. A numerical example of the Analysis of Variance, for the data of Example 13.2, is given in Table 13.4.

EXAMPLE 13.2

Five different aptitude tests, A–E, are applied on five successive days to five different subjects who are considered comparable in intelligence. None of them has previously attempted tests of this type, and so it is required to remove any possible differences between days which could be attributed to a *learning* effect. This is done by using the layout of Fig. 13.2, and the scores obtained are listed in Table 13.4. Investigate the validity of the claim that the tests measure the same qualities, and also examine differences between subjects and between days.

In this layout, rows correspond to subjects and columns to days. The various sums of squares are: Tests, $S_T = \frac{1}{5}(331^2 + 332^2 + 393^2 + 310^2 + 304^2) - 1670^2/25$; Subjects, $S_R = \frac{1}{5}(318^2 + \cdots + 351^2) - 1670^2/25$; Days $S_C = \frac{1}{5}(321^2 + \cdots + 345^2) - 1670^2/25$; and $S = 56^2 + 62^2 + \cdots + 70^2 + 71^2 - 1670^2/25$. These values are shown in Table 13.5, and $S_E = S - S_T - S_R - S_C$ is obtained by subtraction as usual.

Table 13.4   Data of Example 13.2

Scores	Day 1	2	3	4	5	Subject totals	Totals for tests
Subject 1	E: 56	B: 62	A: 65	D: 59	C: 76	318	A: 331
2	C: 74	A: 65	D: 60	E: 61	B: 70	330	B: 332
3	B: 63	E: 59	C: 80	A: 66	D: 64	332	C: 393
4	A: 64	D: 63	B: 67	C: 81	E: 64	339	D: 310
5	D: 64	C: 82	E: 64	B: 70	A: 71	351	E: 304
Day totals	321	331	336	337	345	1670	1670

Table 13.5   Analysis of Variance for Data of Example 13.2

Source of variation	D.F.	S.S.	M.S.	F-ratio
Subjects	4	118·0	29·50	$F_{(4, 12)} = 14\cdot97$ ***
Days	4	62·4	15·60	$F_{(4, 12)} = 7\cdot92$ **
Tests	4	994·0	248·50	$F_{(4, 12)} = 126\cdot14$ ***
Error	12	23·6	1·97	
Total	24	1198·0		

All F values are significant in this analysis. It is unlikely that any reasonable hypothesis could be set up about the way in which subjects differ, so it will be enough to record that they *do* appear to differ, and make no *t*-tests. As for days, there is a steady trend of increasing scores, which will cause no surprise, but the actual daily increases of score are not constant: there seems little point in *t*-tests here either, though perhaps a regression hypothesis (linear or logarithmic) would be worth a trial (as in Chapter 11). For tests A–E, a more detailed examination is now required. Significant differences are $t_{(12)}\sqrt{2\hat{\sigma}^2/5}$; the square root is $\sqrt{2 \times 1\cdot97/5} = \sqrt{0\cdot788} = 0\cdot89$. The values of $t_{(12)}$ at the three standard significance levels are 2·179 (5%), 3·055 (1%) and 4·318 (0·1%), and these values multiplied by 0·89 give the significant differences 1·94 (5%), 2·72 (1%) and 3·84 (0·1%). The mean scores for the various aptitude tests are: A, 66·2; B, 66·4; C, 78·6; D, 62·0; E, 60·8. Application of the significant differences to these means indicates that the tests form three groups: C, which gives the highest scores; A and B, considerably lower than C; D and E, which are somewhat lower than A, B. As usual, it is much more satisfactory to have definite hypotheses to examine, rather than making these comments on looking at the results;

but as the three groups are all separate at the 0·1% level the danger of false conclusions is not very great. Nevertheless, if it had been thought that, say, A and E were much the same, some reflection on the result would be called for, and the structure of these two aptitude tests should then be re-examined.

For Latin Squares to be useful, they must not require an excessive number of replicates (remember that r must equal t); neither must f, the error d.f., be too small. For $t = 4$ we have $f = 6$, and this may be just about adequate where the material being used is not too variable (see Chapter 14). Thus $t = 4$ and 8 are the lower and upper limits of common use of these designs: for $t = 3$, six replicates would be used, laid out as two 3×3 squares.

When choosing a layout for a single experiment, use should be made of Fisher and Yates'[7] tabulation of Latin Squares (Table XV) for $t = 4$, 5 or 6, while for larger sizes it suffices to write a square down out of one's head and then allocate the treatments to letters at random.

Other designs

For some Latin Squares, it is possible to place a further set of treatments (usually denoted by greek letters—hence a *Graeco-Latin Square*) once in each row and column, and once with each of the treatments A, B, C, . . . also. So three systematic sources of variation besides treatments can be taken out. These designs seem to be more readily applicable in industry than in biology, and the reader is referred to textbooks such as that by Cochran and Cox.[3]

In animal experiments, such as dietary trials, genetic differences among the animals (which would be the *unit-plots*) may be so large that any possible way of reducing the effect of these is very desirable. If blocks could be formed from animals of the same litter, all would be well; but very often these blocks prove to be much smaller than the required size of t units. Then designs known as *incomplete blocks*, in which not every treatment need appear in each block, are valuable, the *balanced* incomplete block being the simplest of these. Such designs lack orthogonality, and so are not easy to analyse. The reader is referred to standard textbooks for this topic also.

EXERCISES

(For Answers and Comments, see p. 173)

13.1 Four different plant densities, A–D, are included in an experiment on the growth of lettuce. The experiment is laid out as a randomized block, and the same number of plants is harvested from each plot, giving the weights (kg) recorded below. Examine whether density appears to affect yield. (It may be assumed that planting density increases in the order A, B, C, D.)

Density	Block	I	II	III	IV	V	VI
A		2·7	2·6	3·1	3·0	2·5	3·0
B		3·0	2·8	3·1	3·2	2·8	3·1
C		3·3	3·3	3·5	3·4	3·0	3·2
D		3·2	3·0	3·3	3·2	3·0	3·1

13.2 Six chemical compounds A–F are being tested as potential protective fungicides against apple mildew. Thirty plants of the same variety of apple seedling are available; these are split into 6 groups of 5, and each group of 5 is then dipped into a suspension of one of the compounds A–F. Afterwards the whole collection of plants is placed in a closed environment which is infected with mildew spores. The plants are positioned in 5 randomized blocks to take out any effects due to slight systematic variations in the environment, and subsequently the number of lesions on the first leaf on each plant is counted. These records are tabulated below. Examine the differences between compounds.

Block	I	II	III	IV	V
Compound A	32	22	26	25	18
B	5	8	6	2	4
C	19	12	15	11	14
D	11	3	7	10	8
E	26	16	22	20	13
F	4	3	8	5	1

13.3 A supermarket organisation buys in a particular foodstuff from four suppliers A, B, C, D and subjects samples of this to regular tasting tests by expert panels. Various characteristics are scored, and the total score for the product is recorded. Four tasters *a*, *b*, *c*, *d* at four sessions obtained the results below. Analyse and comment on these.

	Taster			
	a	*b*	*c*	*d*
Session 1	A : 21	B : 17	C : 18	D : 20
2	B : 20	D : 22	A : 23	C : 19
3	C : 20	A : 24	D : 22	B : 19
4	D : 22	C : 21	B : 22	A : 26

14

Precision of Results

When we first considered significance tests, the usual three levels for testing were discussed, and reservations made about their uncritical use. Nevertheless, too much faith is often placed in the very presence or absence of significance between a pair of means under test. This is equally true when comparisons are made between just two sets of data as in Chapter 8, or when the methods of experimental design described in Chapters 12 and 13 are used. Two means may be declared significantly different if the numerical value of their difference exceeds $t_{(f)}\sqrt{2\sigma^2/r}$, the *significant difference* defined in Chapter 12; so that an experiment may be said to have good *precision* if $t_{(f)}\sqrt{2\sigma^2/r}$ is relatively small, and to have poor precision (to be *insensitive*) if this expression is large. An insensitive experiment is thus one in which only very large differences between means can be shown significant statistically, and is hardly likely to add to our knowledge about the treatments under study.

Precision in this sense can be improved by reducing the numerical values of some or all of the quantities in the expression $t_{(f)}\sqrt{2\sigma^2/r}$. The effect of *replication* is twofold. First, it improves precision to increase r, for the expression contains $1/\sqrt{r}$ and so can be, for example, halved by quadrupling r. Second, an increase in r helps to increase f, the error d.f.; f also depends on whether the design is completely randomized, randomized blocks, or some other suitably chosen one. In its turn, f has two effects: one is to provide a well-based estimate of σ^2 (to which confidence-limits, if applied as in Chapter 9, would be reasonably narrow) and the other is to ensure that the values of $t_{(f)}$ to be used in significant differences are not too high. The latter is achieved if $f = 10$ or more, for a glance down the column headed $P = 0.05$ in the t-table (Table I) shows that the numerical value of $t_{(0.05, f)}$ increases quite rapidly as f decreases below 10 while it alters relatively little, and slowly, for f greater than 10. The amount of replication needed to give $f = 10$ or more depends on the type of design used and the number of treatments included; in theory, if there is a large number of treatments r need not be more than 2 (it must be at least 2, to give any replication on which to base σ^2), but in practice most workers would not be content with means based on less than 4 replicates.

However impractical, it might appear at first sight that 'the more the better' is the slogan for replication: but this is no wiser than most slogans, for the increase in precision obtained by raising f above 10 and r above 4 (or whatever its value must be so that f shall be 10) is often more than offset by an increase in σ^2. This can be due to several causes. It may no longer be possible to find a sufficient number of homogeneous units of material; a field experiment may have to go into two patches of land which have not had the same previous treatment, or different sources of seed may have to be used to sow the field. The experiment may become so unwieldy that inaccuracies creep into the recording, due either to fatigue or to pressure of time, or in some laboratory trials the last few samples examined may have deteriorated through being kept too long. Additional observers may have to take part in the recording, the chemical analysis of plant parts, the examination of microscope slides and so on, and this sometimes causes extra sources of systematic variation which, if they are not designed out, will inflate the value of σ^2.

The actual value of σ^2 is clearly important: it can be minimized by making sure that all systematic variation is allowed for (using, if needed, refinements of design like those mentioned at the end of Chapter 13), by choosing plot sizes so that errors in recording are small in relation to the actual size of the measurement made (so that ten blackcurrant bushes or twenty strawberry plants are commonly used as unit-plots in field experiments, giving a total crop per plot which is hardly affected by such recording errors as are reasonable), and also by a method known as the analysis of covariance, for which reference should be made to the text by Cochran and Cox.[3] In greenhouse or laboratory work, conditions can sometimes be controlled sufficiently for σ^2 to be substantially smaller than could ever be expected in field experiments, and some relaxation of the lower limits to r and f may be permitted.

When σ^2 is known roughly, or can be guessed, from previous trials on the same or very similar crops, it is possible to say how many replicates will be needed to achieve a given degree of precision. Let us express the precision required by saying that if a pair of means differ numerically by an amount δ or more, we wish to show that they are statistically significantly different. This simply says that the significant difference $t_{(f)}\sqrt{2\sigma^2/r}$ must be less than (or equal to) δ.

EXAMPLE 14.1

Suppose that σ^2 is approximately 25, and δ is required to be 6. Find the least value of r which will achieve this.

Unless r is known, f is not known exactly, so for a first approximation $t_{(f)}$ is replaced by 2. Then $2\sqrt{50/r} \leqslant 6$, so that $\sqrt{r} \geqslant \frac{2}{6}\sqrt{50}$ or $r \geqslant 50/9$, and since r has to be a whole number this gives $r = 6$ as the minimum replication needed to achieve the stated aim. It is possible that when the design has been chosen and the number of treatments settled, so that f and $t_{(f)}$ can be quoted exactly, r may need increasing to the next whole number. But the stated aims are only approximate, and we do not in any case know whether the estimate of σ^2 from this particular experiment will be very near to 25, since the value 25 can only be inferred from previous experience with the same type of experimental unit and the variance is bound to change from one experiment to another. This method should be regarded as giving an idea of what order of replication is needed—whether r must be large, say 10 or 12, or a more reasonable value such as 4 or 5. It may lead to an inconveniently large value of r, and so to a reconsideration of whether the stated aim for precision is necessary. And it may lead also to a decision that an experiment with the resources available is just not worth doing for the value of information likely to come from it—one suspects that this decision is not taken nearly as often as it should be.

EXAMPLE 14.2

Sometimes we are interested in detecting whether a treatment changes a mean *by some stated percentage*: for example, can we increase the yield of a crop by *at least* 7% by adding an extra fertilizer element? The idea here is that unless we can achieve at least a 7% increase it will not pay the cost of the extra fertilizer. So the requirement of precision in this situation may be expressed as the ratio

$$\frac{\text{increase in crop yield by using new treatment}}{\text{average crop yield under standard treatment}}.$$

In order to carry out the analysis in this type of problem we need not simply an estimate (or guess) of σ^2, but instead the ratio of σ to the mean yield of the standard treatment. This ratio σ/mean is called the *coefficient of variation*, V. Suppose it is 16% in our example: that is to say we are told δ/mean $= 7\%$ and σ/mean $= 16\%$. Thus $t_{(f)}\sigma\sqrt{2/r} \leqslant \delta$ gives $t_{(f)}\sqrt{2/r} \times 16 \leqslant 7$, or approximately $32\sqrt{2} \leqslant 7\sqrt{r}$, giving $r = 42$ approximately. So we are asking for too high a level of precision to be reasonable, with material of this degree of variability.

It is not unusual in field experiments for the coefficient of variation, V, to reach 20%, but in laboratory and greenhouse studies something like $V = 5$ to 10% should usually be attainable, so that to achieve precision equivalent to detecting a difference $|\delta| = 10\%$ should be reasonable in the latter case: a calculation like the previous one suggests a quite practicable value for r. Using this r, we would expect that if the mean yield under the standard treatment was 100, and a new treatment could raise this to at least 110, the difference between these two treatments would usually show up as significant.

EXERCISES

(For Answers and Comments, see p. 173)

14.1 The mean height to which wheat plants grow in a fixed period of time is 18 cm, and the standard deviation of height is 2·8 cm. An experiment using 8 treatments is laid out in 4 randomized blocks. What is the least difference in mean height between two treatments that can be shown as significant in this experiment?

14.2 In a field experiment using plots of 4 m², it is known from past experience that the coefficient of variation of crop yield will be about 8%. What is the minimum replication that will enable a 10% difference in treatment mean yields to be detected in a significance test?

If plots of 2 m² are used instead, so that this replication can be doubled, the coefficient of variation will go up to 12%. What difference can now be detected as significant?

15

Simple Factorial Experiments

So far we have not considered that the treatments in an experiment may be related to one another; all the designs described would, for example, suit perfectly well for comparing a variety of fungicides which are chemically quite different in structure, and have only one similarity, namely that they all act against the same fungus upon which the experiment is being conducted. But it is often necessary to make up a set of treatments from basic *factors*, for example in fertilizer trials. A fertilizer needs to contain the elements N, P, K, Mg (nitrogen, phosphorus, potassium, magnesium), and the treatments examined in an experiment will be various combinations of amounts of these elements. However, the real topic of interest is not these particular combinations but the basic factors, the elements themselves: the most interesting question of all is whether the factors act independently of one another. If we consider how much N to apply, and then how much P, it is very likely that the best amount of P will depend very closely on how much N is present (whereas if fungicides are included in the same experiment, to keep plants *clean*, it is much less likely that the N or P requirement of a plant would be affected by which fungicide had been used on it).

As an illustration of the fundamental idea of **interaction**, suppose that a fertilizer containing only basic amounts of N and P induces an average yield of 5 units in a crop; and that this can be increased to 30 units or 20 units respectively by adding extra N or P respectively, *without* the other. The response to extra N is thus 25 units (30 minus 5) and to P is 15 units (20 minus 5). But this gives an incomplete picture until we know what happens when the extra amounts of N and P are both added *together*. If N and P act independently, the response to N and P together will be the sum of the two responses already measured, namely 25 plus 15, i.e. 40 units, so that the actual yield (basic plus response) is 45 units. However, we are very likely to find that the response to N can be enhanced by adding P as well, so that with both together the yield proves to be *more* than 45 units: we then say that N and P *interact*, and in this case the interaction is a positive one. Some combinations of elements can result in a depression of the predicted yield, which would be called a negative interaction, when they are present in quantities that give the plants an unbalanced nutrient regime: the elements K and Mg in fertilizers can give this effect, and even be toxic when the unbalance is severe.

Whenever it seems likely that two or more basic elements or factors will interact, little or no useful information is obtained by experimenting on them one at a time. They must be examined together, in a design known as a ***factorial experiment***, which provides information on how the factors interact, either in pairs (NP, NK, etc.) or larger groups (NPK, etc., wherein the extent of interaction between N and P can be altered by altering the amount of K used, and so on).

Notation

There is a standard notation for factorial experiments, which unfortunately clashes to some extent with that traditionally used for experiments in general, as described in earlier chapters. In factorial experiments the factors are denoted by capital letters A, B, C, . . . ; the actual treatments used in the conduct of a factorial experiment will consist of combinations of various amounts (*levels*, as they are called) of these factors, and we can no longer denote the treatments by capitals.

EXPERIMENTS WITH 2-LEVEL FACTORS

A useful type of factorial experiment, which can be employed at the outset of a programme of work to discover just which out of a whole set of possible factors do interact with one another, has each factor included at two levels only: high and low, or present and absent. If, for example, there are just three likely factors to be investigated, we shall call them A, B and C; and the first experiment in a series to study their behaviour will contain a high and low level of each. These levels will be used in all possible combinations, of which there are $2 \times 2 \times 2$; there is a standard method of denoting these, as follows. Let the symbol *abc* stand for that combination which has A, B, C all present (or all at the high level); *ab* has A, B present and C absent (or A, B at the high level and C at low); *bc* has B, C present and A absent; *ac* has A, C

present and B absent; *a* has A present, B and C absent; *b* has B present, A and C absent; *c* has C present, A and B absent; ① has all three absent.*
These $2 \times 2 \times 2$, or 2^3, that is 8, treatment combinations will then be laid out in an experimental design appropriate to the conditions under which the work is to be done (e.g. in randomized blocks, replicated a reasonable number of times). It is only when we come to consider the 7 degrees of freedom for treatments that we alter the construction of the analysis of variance so as to show the effects of, and interactions between, factors. These 7 d.f. correspond to the sum of squares between the totals of the eight treatment combinations ①, *a*, *b*, *c*, *ab*, *ac*, *bc*, *abc*. It is possible to define seven single degrees of freedom: three for **main effects**, denoted by the capital letters A, B, C; three for the *first-order interactions* between pairs of elements, denoted by AB, AC, BC; and one for the **second-order interaction** of all three elements, ABC. Note here again that the capital letters used to denote the factors in an experiment are also used to denote the main effects, but *not* the actual combinations of factor levels that make up the treatments applied to the experimental plots.

The main effect of A is the average response to addition of A, averaged over all the combinations of other factors (B, C) in the experiment. It is thus $\frac{1}{4}[(abc - bc) + (ab - b) + (ac - c) + (a - ①)]$, since each bracket ( ) represents a response to A, i.e. a difference between a plot which contains A and one which does not, both having the same levels of B and C; there are four such brackets to be averaged. Obviously this is equivalent to taking the difference between all plots with A and all without, so that

$$A = \frac{1}{4}[(abc + ab + ac + a) - (bc + b + c + ①)], \quad \text{and similarly}$$
$$B = \frac{1}{4}[(abc + ab + bc + b) - (ac + a + c + ①)],$$
$$C = \frac{1}{4}[(abc + ac + bc + c) - (ab + a + b + ①)].$$

The interaction AB is defined as the average response to A in the presence of B minus the average response to A in the absence of B, averages in each case being taken over all possible combinations of the remaining factors in the experiment (in this example, just the presence and absence of C). Thus we obtain

$$AB = \frac{1}{4}[\{(abc - bc) + (ab - b)\} - \{(ac - c) + (a - ①)\}]$$

i.e.

$$AB = \frac{1}{4}[(abc + ab + c + ①) - (bc + b + ac + a)]$$

AC and BC may be obtained by similar arguments, but the expressions for interactions are not easy to remember, and fortunately there exists a convenient algebraic way of finding out what expression to use for

* The symbol ① here denotes what is usually written (1).

any main effect or interaction of any order; it gives the following results.

$$\textit{Main effects:} \quad \begin{aligned} A &= \tfrac{1}{4}(a-1)(b+1)(c+1) \\ B &= \tfrac{1}{4}(a+1)(b-1)(c+1) \\ C &= \tfrac{1}{4}(a+1)(b+1)(c-1) \end{aligned}$$

$$\textit{First-order interactions:} \quad \begin{aligned} AB &= \tfrac{1}{4}(a-1)(b-1)(c+1) \\ AC &= \tfrac{1}{4}(a-1)(b+1)(c-1) \\ BC &= \tfrac{1}{4}(a+1)(b-1)(c-1) \end{aligned}$$

$$\textit{Second-order interaction:} \quad ABC = \tfrac{1}{4}(a-1)(b-1)(c-1)$$

Write down the brackets $(a \quad 1)$ etc. for every factor used in the experiment, and whenever a capital letter appears in the effect or interaction being calculated, put a minus sign into the corresponding bracket. Plus signs go into the other brackets and have the effect of averaging over the factors represented by these brackets. The expressions are evaluated just as though they were simple algebraic products, e.g.

$$AB = \tfrac{1}{4}(a-1)(bc+b-c-1) = \tfrac{1}{4}(abc+ab-ac-a-bc-b+c+①)$$

as in the definition. Although verbal definitions for higher-order interactions are complicated, these too can be calculated in this way, and we obtain

$$ABC = \tfrac{1}{4}(abc-ab-ac-bc+a+b+c-①)$$

Whenever interactions are considered possible, factorial designs must be used; but if interactions are not found, nothing is lost, for the main effects still represent the results of a well-replicated trial on each element separately, comparing, for A, just the two levels *present* and *absent*, each level based on half the total number of plots in the whole experiment.

After preliminary 2^n experiments (n factors each at 2 levels), those factors that prove of interest will need much closer study, and 3-level experiments, which are common, or higher numbers of levels of a few factors, are needed. We give a simple example of such an experiment later in this chapter, but the general theory of these, and especially how to proceed when the number of treatment combinations is very large, is not easy and readers should consult a more advanced text such as that by Cochran and Cox.[3]

Example of a 2^3 experiment

EXAMPLE 15.1

Table 15.1 summarizes the increases in weight of guinea-pigs fed on basic or supplemented diets; three different supplementing factors, bran, vitamins and antibiotics, were being examined. Each diet was fed to four animals, one in each of four pens; thus the experimental layout was randomized blocks, with pens forming blocks.

Table 15.1

Diet	Basic	+Anti-biotics (A)	+Bran (B)	+Vita-mins (V)	+A and B	+A and V	+B and V	+A, B and V	Pen totals
	①	a	b	v	ab	av	bv	abv	
Pen I	7	8	11	10	12	12	15	16	91
II	8	10	11	11	15	16	16	15	102
III	6	8	13	12	14	12	13	15	93
IV	7	8	12	10	14	14	17	16	98
	28	34	47	43	55	54	61	62	384

Table 15.2 gives the analysis of the layout as it was, in randomized blocks, ignoring the factorial nature of the treatments: from it we obtain an estimate of σ^2 in the usual way (from the error-mean-square)

Table 15.2   First-stage Analysis of Variance

Source of variation	D.F.	S.S.	M.S.	
Pens	3	9·250	3·0833	$F_{(3, 21)} = 2·42$ n.s.
Treatments (Diets)	7	268·000	38·2857	
Error	21	26·750	1·2738	
Total	31	304·000		

and also a test of whether segregation into pens removed any significant amount of variation, which in the example it did not. Seven single d.f. are now obtained; in the definition of any particular effect or interaction there are four treatments carrying a plus sign and four a minus, e.g. for A, each of abv, ab, av, a appears with a plus and b, v, bv, ① with a minus. The total for A is thus $62 + 55 + 54 + 34 - 47 - 43 - 61 - 28 = +205 - 179 = +26$, and the single d.f. for A has value $(+26)^2/32$, 32 being the complete number of experimental plots (animals). Similarly the total for B is $+225 - 159 = +66$, for V $+220 - 164 = +56$, for AB $+188 - 196 = -8$, for AV $+191 - 193 = -2$, for BV $+185 -$

$199 = -14$, and for ABV $+186 - 198 = -12$. Squaring each of these and dividing by 32, we obtain the single d.f. listed in Table 15.3, and as a check they do indeed add to the 7 d.f. *treatments* term already obtained. In this method, the factor $\frac{1}{4}$ in the original definition is not used (though of course it is included in the divisor 32).

Table 15.3   Full Analysis of Variance

Source of variation	D.F.	S.S.	M.S.	
Pens	3	9·250	3·0833	$F_{(3, 21)}$ = 2·42 n.s.
A	1	21·125	21·1250	$F_{(1, 21)}$ = 16·58 ***
B	1	136·125	136·1250	$F_{(1, 21)}$ = 106·86 ****
V	1	98·000	98·0000	$F_{(1, 21)}$ = 76·93 ***
AB	1	2·000	2·0000	$F_{(1, 21)}$ = 1·57 n.s.
AV	1	0·125	0·1250	F < 1 n.s.
BV	1	6·125	6·1250	$F_{(1, 21)}$ = 4·81 *
ABV	1	4·500	4·5000	$F_{(1, 21)}$ = 3·53 n.s.
Treatments	7	268·000		
Error	21	26·750	1·2738	
Total	31	304·000		

We now see that each of the main-effect terms has a very large value of F, significant at 0·1%; but on looking at the rest of the table we discover an interaction, BV, which is significant. Thus, while it is correct to examine the response to A in terms of its main effect, the responses to B and V have to be considered together, and it would be misleading to say anything specific about their main effects. If we consider A first, the total weight increase of the 16 animals receiving A was 205, so that the mean per animal was 12·8; the mean increase without A was similarly $179/16 = 11·2$, and so the response to A is a beneficial one. For B and V, we need a *two-way* table of means: 8 animals received neither B nor V, their total weight increase being 62 (*a* and ①), and the mean thus 7·75; the mean with B but not with V (*b* and *ab*) is 12·75, with V but not B (*v* and *av*) 12·13, and with both B and V (*bv* and *abv*) 15·38. The significant differences between any two of these means based on 8 observations are $t_{(21)}\sqrt{2\hat{\sigma}^2/8} = t_{(21)}\sqrt{\frac{1}{4} \times 1·2738} = 0·564 \times t_{(21)}$. Putting in the tabular values of $t_{(21)}$, namely 2·080 (5%), 2·831 (1%), 3·819 (0·1%), we obtain the significant differences 1·17 (5%), 1·60 (1%) and 2·15 (0·1%). A table of means, Table 15.4, is now written out. The B-effect is significant at 0·1% both in the presence of V and in its absence, as is seen by looking at the two rows of the table in turn;

Table 15.4   Two-way table of means for B and V

	− B	+ B
− V	7·75	12·75
+ V	12·13	15·38

similarly, the V-effect is significant at 0·1% for either − B or + B, and
the only reason for an interaction proving significant here seems to be
that the sizes of the B- and V- effects, although always very significant,
vary somewhat numerically.

When a 2-level factorial experiment has indicated which single factors or groups of factors have a significant effect on the records that
have been analysed, a more detailed study of the form of the effect is
required. In Example 15.1, B and V must be examined together in a
further experiment wherein each factor is present at three levels at
least, though since A did not appear to affect either of them there would
be no need to study it at the same time.

A single factor (such as A above) showing a main effect in a 2-level
experiment requires to be studied further at several levels, with a view
to establishing a definite equation relating its level to the response y.
As an example, consider the amount of nitrogen present in a fertilizer
applied to plants; suppose that the two points marked × in Fig. 15.1

Fig. 15.1   Effect of amount of nitrogen fertilizer on crop yield; two points ×
obtained in a preliminary experiment, and three points ◯ in later detailed study.

represent the levels used in the original experiment. There are many
forms which a curve relating yield y to applied level N might take be-

tween these two points. A reliable curve can hardly be based on less than five observations, so that in further experiments at least the three points marked ◯ should be added, giving a minimum of five levels of N studied simultaneously. Curves similar to that in Fig. 15.1 are common in studies of plant growth; on occasion also something very near to a straight line may be obtained, or even a curve which has a maximum yield at an intermediate level of N and begins to fall again at the higher levels investigated.

If two factors, say N and P in a fertilizer, interact, we expect to find a different form of response (crop yield, etc.) to P according to whether N has a higher or lower level. Let us suppose that N is present at three levels and P at five; then Fig. 15.2 shows the type of response curve

Fig. 15.2 Typical responses to phosphate fertilizer when nitrogen is present at three different levels N_1, N_2, N_3.

which may arise. When more than two factors are involved, it becomes progressively more difficult to interpret the situation from diagrams, and it is necessary to fit an equation to the data. The examination of growth curves in general in biology is an extensive subject, and some of the methods of proceeding are explained by Cochran and Cox.[3]

Example of a mixed factorial

When factors are not all used at the same number of levels, we call the experimental design a *mixed factorial*. We give below the analysis for a design using three fertilizers N, P and K on strawberries, in which N was included at two levels and each of the other factors at three levels. This is a '2 × 3 × 3' experiment, because there are 2 × 3 × 3 = 18 different combinations of levels of N, P, K that we can think of; these 18 combinations will form the treatments in the experiment. Factors at more than two levels need a more general notation than we have so far had to use: we shall call the three levels of P P_1, P_2, P_3 and those of K K_1, K_2, K_3. To avoid confusion we shall write the two levels of N in the same notation as N_1, N_2.

EXAMPLE 15.2

The amount of nitrogen (ppm) present in dried leaves of strawberries was measured from an experiment having two levels of N and three each of P and K, the full set of treatments being replicated twice in randomized blocks. Table 15.5 gives the list of treatment combinations and the detailed records of the leaf analysis. Treatment $N_1P_1K_1$ denotes the lowest level of each factor applied to the plants, $N_2P_3K_3$ the highest level of each.

In the Analysis of Variance, we expect, as before, to be able to set up

Table 15.5   Leaf nitrogen (ppm) present in dried strawberry leaves from 2 x 3 x 3 N x P x K fertilizer trial

Treatment	Block I	Block II	Total
$N_1P_1K_1$	2·42	2·45	4·87
$N_1P_1K_2$	2·46	2·44	4·90
$N_1P_1K_3$	2·48	2·51	4·99
$N_1P_2K_1$	2·43	2·44	4·87
$N_1P_2K_2$	2·48	2·46	4·94
$N_1P_2K_3$	2·52	2·51	5·03
$N_1P_3K_1$	2·47	2·50	4·97
$N_1P_3K_2$	2·50	2·53	5·03
$N_1P_3K_3$	2·54	2·55	5·09
		Total for N_1:	44·69

Treatment	Block I	Block II	Total
$N_2P_1K_1$	2·48	2·50	4·98
$N_2P_1K_2$	2·47	2·52	4·99
$N_2P_1K_3$	2·53	2·55	5·08
$N_2P_2K_1$	2·47	2·51	4·98
$N_2P_2K_2$	2·52	2·50	5·02
$N_2P_2K_3$	2·55	2·52	5·07
$N_2P_3K_1$	2·50	2·54	5·04
$N_2P_3K_2$	2·56	2·51	5·07
$N_2P_3K_3$	2·57	2·55	5·12
		Total for N_2:	45·35

Total for Block I = 44·95, for Block II = 45·09. Grand Total G = 90·04

main-effect terms for each of the factors N, P, K and also terms for interactions between any pair of these and between all three. There are only
two levels of N, and so the main effect of N will have 1 degree of freedom
as before. But P and K have each been included at three levels; so for
each of these factors the main effect must measure the difference between
three sets of plots—those receiving level 1, those receiving 2 and those
receiving 3. We noted in Chapter 12 that if there are t treatments being
compared, $(t-1)$ d.f. will be used in making the comparison. In the same
way, in a factorial experiment, the d.f. for the main effect of any factor will
be one less than the number of levels of that factor. P and K in this example
both have 2 d.f. The rule for interactions is that d.f. for NP = (d.f. for N) ×
(d.f. for P), and similarly for NK, PK; and in the same way d.f. for
NPK = (d.f. for N) × (d.f. for P) × (d.f. for K). Therefore in this example
NP and NK each have 2 d.f., while PK and NPK each have 4 d.f. Now
that we have assigned degrees of freedom to all the possible main effects
and interactions of the three factors, we add these d.f. together and check
that they do account for all the 17 d.f. between the 18 treatment combinations used in the experiment.

In order to calculate main effects we shall need the totals level-by-level;
for interactions we shall need two-way tables. So we extract from Table
15.5 three two-way tables, which are shown in Table 15.6. In the table

Table 15.6 Two-way tables showing totals summed over all blocks and all levels
of the third factor

	P_1	P_2	P_3		N totals
N_1	14·76	14·84	15·09	:	44·69
N_2	15·05	15·07	15·23	:	45·35
P totals:	29·81	29·91	30·32		90·04 $= G$

	K_1	K_2	K_3		N totals
N_1	14·71	14·87	15·11	:	44·69
N_2	15·00	15·08	15·27	:	45·35
K totals:	29·71	29·95	30·38		90·04 $= G$

	K_1	K_2	K_3		P totals
P_1	9·85	9·89	10·07	:	29·81
P_2	9·85	9·96	10·10	:	29·91
P_3	10·01	10·10	10·21	:	30·32
K totals:	29·71	29·95	30·38		90·04 $= G$

for N and P, the entry in the first row and first column is the sum of *all* plots that had the lower level 1 of N and the lowest level 1 of P, from both blocks and all three levels of K: it is thus the sum of 6 plot measurements. The other five entries in the table are made up in the same way. In the margins of the table are shown the overall totals for N_1 and N_2 (rows) and also for P_1, P_2 and P_3 (columns). The NK table is made up in just the same way; so is the PK table, except that each entry here is the sum of 4 plot measurements (from both blocks and two levels of N).

The analogy between d.f. for t treatments and d.f. for a main effect gives us the key to computing the sum of squares for a main effect: we simply use the totals for each level and calculate the sum of squares between them. Thus for P, the totals are 29·81 (P_1), 29·91 (P_2) and 30·32 (P_3), which add to $G = 90·04$. N, the total number of plots, is 36, and each P total is based on 12 plots. The sum of squares for the main effect of P is thus $\frac{1}{12}(29·81^2 + 29·91^2 + 30·32^2) - 90·04^2/36$. That for K is $\frac{1}{12}(29·71^2 + 29·95^2 + 30·38^2) - 90·04^2/36$ and, for N, $\frac{1}{18}(44·69^2 + 45·35^2) - 90·04^2/36$, since each N total is based on 18 plots. The values of all these are shown in the Analysis of Variance, Table 15.7.

Turning to interactions between two factors, we ask what the NP two-way table of totals, in Table 15.6, can tell us besides the information about main effects of N and P that we have already extracted from the margins. The situation is exactly like that in a 2×2 table: if the entries in the body of the table can be predicted as soon as we know the margins, then the factors do not interact, but if (for example) the pattern of P response under N_1 is very different from that under N_2 we need to be given all the separate

Table 15.7   Analysis of Variance for Example 15.2

Source of variation	D.F.		S.S.	M.S.	
Blocks	1		0·00054	0·00054	$F_{(1,17)} = 1·29$ n.s.
N	1	0·01210		0·01210	$F_{(1,17)} = 28·81$***
P	2	0·01217		0·00609	$F_{(2,17)} = 14·50$***
K	2	0·01921		0·00961	$F_{(2,17)} = 22·88$***
NP	2	0·00095		0·00048	$F_{(2,17)} = 1·14$ n.s.
NK	2	0·00071		0·00036	$F_{(2,17)} < 1$ n.s.
PK	4	0·00053		0·00013	$F_{(4,17)} < 1$ n.s.
NPK	4	0·00019		0·00005	$F_{(4,17)} < 1$ n.s.
Treatments	17		0·04586		
Error	17		0·00716	0·00042	
Total	35		0·05356		

NP totals—the margins are not enough—and there *is* an interaction between N and P. The six NP totals here contain the information on the N main effect (when added by rows), the P main effect (when added by columns) and, in addition, the NP interaction. There are 6 totals, so 5 d.f. for comparing them; 1 d.f. is the N main effect, 2 d.f. are the P main effect, and the remaining 2 d.f. are the NP interaction. The 5 d.f. sum of squares from these six totals is $\frac{1}{6}(14 \cdot 76^2 + 15 \cdot 05^2 + 14 \cdot 84^2 + 15 \cdot 07^2 + 15 \cdot 09^2 + 15 \cdot 23^2) - 90 \cdot 04^2/36 = 0 \cdot 02522$, each total being based on six observations; this is $N + P + NP$, from which we easily compute NP. Similar calculations from the other two tables in Table 15.6 give NK and PK, whose values are shown in Table 15.7.

It now only remains to calculate the 3-factor interaction NPK. The best way to do this is to work out the total sum of squares for the whole set of 18 treatment combinations used in the experiment, and to subtract from that the sums of squares for each main effect and two-factor interaction already found. The 18 treatment totals (each one the sum of just two entries, from Block I and Block II) are shown in Table 15.5, and give the sum of squares $\frac{1}{2}(4 \cdot 87^2 + 4 \cdot 90^2 + \ldots + 5 \cdot 07^2 + 5 \cdot 12^2) - 90 \cdot 04^2/36 = 0 \cdot 04586$. The Analysis of Variance is completed by working out the total sum of squares for all 36 plots, $(2 \cdot 42^2 + 2 \cdot 45^2 + \ldots + 2 \cdot 57^2 + 2 \cdot 55^2) - 90 \cdot 04^2/36$ and the sum of squares for blocks.

From Table 15.7, we see that there is no evidence of any interaction, the only significant terms in the analysis being the three main effects. At once it is clear that N_2 gives higher results than N_1, since there are only two N levels to compare. But there are three P levels, so we require the three means: P_1 has mean $29 \cdot 81/12 = 2 \cdot 484$, P_2 has mean $2 \cdot 493$ and P_3 has mean $2 \cdot 527$. Each of these means is based on 12 observations, and so we may compare them by working out the least significant difference $t_{(17)} \sqrt{2\hat{\sigma}^2/12} = t_{(17)} \sqrt{0 \cdot 00042/6} = 0 \cdot 00837 \, t_{(17)}$. The values of $t_{(17)}$ from Table I are $2 \cdot 110$ (5%), $2 \cdot 898$ (1%), $3 \cdot 965$ (0·1%) and so significant differences are $0 \cdot 018$ (5%), $0 \cdot 024$ (1%) and $0 \cdot 033$ (0·1%). From these we see that P_1 and P_2 do not differ significantly in mean, but P_3 is greater than either of these at the 0·1% significance level. The means for K are also based on 12 observations each, and so we can use the same significant differences; K_1 mean $= 2 \cdot 476$, K_2 mean $= 2 \cdot 496$, K_3 mean $= 2 \cdot 532$. This time K_1 is below K_2 at the 5% significance level, and K_3 is greater than either at the 0·1% significance level. This completes the summary of results for this Example, but if there had been any significant interactions in the Analysis of Variance we would have dealt with them as in Example 15.1 (see Table 15.4 and the explanation of how to test the means in that two-way table).

EXERCISES

(For Answers and Comments, see p. 174)

15.1 A fertilizer trial on strawberries consists of 4 replicates of the 4 treatments ①, n, p, np, combinations of high and low levels of nitrogen and phosphorus. The resulting crop yields per plot (in suitable units) are:

	①	n	p	np
Block I	13	24	16	27
II	12	25	14	34
III	18	24	15	32
IV	15	31	20	30

Carry out an analysis to determine the effects (if any) due to blocks, N, P and the interaction NP.

15.2 Two spray materials A, B, are applied to apple trees; the effect of A is to reduce damage to fruit due to scab, and B controls mildew on shoots but may damage the skin of the fruit. Four treatments (A alone, B alone, both together, neither) are employed in 5 replicates, and the appearance of a sample of fruit assessed on a scoring scale with the following results (low scores imply poor appearance):

	Neither A nor B	A alone	B alone	Both A and B
Block I	12	24	10	12
II	15	28	12	15
III	8	23	10	19
IV	17	20	14	11
V	13	30	9	16

Analyse and comment on these results.

15.3 Seedling plants are transplanted from a greenhouse to an outdoor plantation at three different times, T_1, T_2 and T_3, being given one of four different hormone rooting treatments R_1, R_2, R_3, R_4 when transplanted. Two plants are used for each RT combination, the layout being completely randomized. The results are summarized in the following table: the entry 11; 13 for the R_1T_1 combination indicates that one plant grew 11 cm and the other 13 cm in a fixed period of time after transplanting; and so on.

	T_1	T_2	T_3
R_1	11;13	18;12	18;17
R_2	15;21	20;23	15;16
R_3	18;18	21;20	19;20
R_4	19;16	19;20	26;24

Analyse and comment on these results.

16

Missing Observations and Non-Normality

MISSING OBSERVATIONS

When an experiment has been designed according to statistical prin-ciples, it may be analysed by one of the relatively simple methods already described. But these methods are based on totals and means of groups of observations; thus they cannot be applied when some of the plots in the layout cannot be recorded. If this happens, the totals are incomplete, in the sense that (for instance) not every treatment is represented in every block total if the layout was randomized blocks. At the end of Chapter 13 it was pointed out that incomplete block designs are much less easy to analyse than the orthogonal ones considered in this book. Fortunately, as we shall see below, it is fairly easy to deal with the difficulties which arise when only one or two plots in an orthogonal layout cannot be recorded: these we describe as *missing plots* or *missing observations*. If one of the treatments in a randomized block fails completely it may simply be omitted from the analysis, since it affects every block equally; and if one whole block is lost, similar omission is made.

Often, one plot may be lost for a reason not attributable to treatment: plants in a field or greenhouse experiment on fertilizers may become diseased; there may be accidental damage by cultivation machinery; an analytical sample or a microscope slide may be lost or damaged in a labora-tory; an animal in a nutrition trial may become ill through disease. Before we treat a lost observation as a 'missing value' we must be quite sure that the loss has not been caused by the experimental treatment: if in a fertilizer experiment plants die on plots which have received a fertilizer that lacked some essential nutrient element, the value o is the correct record of crop yield; in this case much useful information might be obtained by recording features other than yield, such as the heights to which plants have grown, numbers of leaves formed, and so on—records which would *not* read o for these plots and would help to pinpoint the effect of fertilizer deficiency.

We will assume in what follows that only one plot has been lost and that it is clear beyond reasonable doubt that the loss has been accidental or independent of treatment. The solution in this case follows these three steps:

1 a value is estimated for the observation on the lost plot;
2 this estimate is written into the table of observed records and the analysis now carried out by exactly the same method as usual, i.e. as described in Chapter 13, for the design which has been used; *except* that
3 the degrees of freedom for total and error sums of squares are each reduced by 1.

The required estimate of the missing observation depends on the type of design used.

Randomized blocks

The table of observations is written out as in Example 13.1, and has one blank space where the observation has been lost; this will affect one particular block and one particular treatment, whose totals are incomplete because they are calculated from one observation fewer than they should be. Call these incomplete totals B' and T' respectively; the grand total is of course similarly incomplete, and we label it G'. The estimate, x, of the lost observation uses just these incomplete totals, as follows:

$$x = \frac{rB' + tT' - G'}{(r-1)(t-1)}$$

This value of x is written into the blank space, totals recalculated, and the analysis continued as normal except that the total sum of squares has $(rt-2)$ rather than $(rt-1)$ d.f., and the error d.f. are also reduced by 1.

EXAMPLE 16.1

Four diets, A–D, were compared in a specially-bred strain of guinea-pigs. Six litters, I–VI, were available, and were used as *blocks* to take out genetic variation. Increases in weight over a period are given in Table 16.1 in suitable units; the animal from litter IV which should have received diet D showed signs of disease and was removed from the experiment before the final recording.

Table 16.1   Increases in weights of guinea-pigs: data of Example 16.1

Litter	I	II	III	IV	V	VI	
Diet A	24	34	41	27	36	32	
B	35	38	46	33	37	35	
C	40	44	54	38	46	40	
D	29	35	40		38	34	$T' = 176$
			$B' = 98$				$G' = 856$

In this experiment $r = 6$, $t = 4$. Thus

$$x = \frac{6 \times 98 + 4 \times 176 - 856}{5 \times 3} = \frac{436}{15} = 29 \cdot 07$$

Since all the other measurements are whole numbers, x will be taken to the nearest whole number, 29; this value is entered in the empty space in the table, and the analysis now carried out as in Example 13.1.

Latin squares

Here a single lost observation will affect one row, one column and one treatment: the incomplete totals of these are written R', C', T' and the incomplete grand total G'. Then

$$x = \frac{t(R' + C' + T') - 2G'}{(t-1)(t-2)}$$

A missing plot estimation does not in any sense *recover* the information which has been lost; its purpose is only to allow the simple form of analysis, characteristic of orthogonal designs, to be maintained with the minimum of adjustment. Adjustment of the degrees of freedom has already been mentioned, and in the expression for a significant difference $t_{(f)}$ must have the adjusted value for f, since the d.f. for t are always those for $\hat{\sigma}^2$. The square root $\sqrt{2\hat{\sigma}^2/r}$ in the expression for significant difference will not need to be adjusted as long as two unaffected treatments, such as A and B in Example 16.1, are being compared, but when the treatment which suffers from a missing observation (D in the example) is being compared with any of the others we use

$$\sqrt{\hat{\sigma}^2 \left(\frac{1}{r} + \frac{1}{r-1} \right)}$$

instead. A warning should be given that the significance levels of the F-and t-tests made when observations are missing do not prove to be exactly 5%, 1% or 0·1%, but something very near to these, and this mathematical disturbance can be neglected so long as the results of significance tests are not interpreted too rigidly.

When more than one plot is lost in randomized blocks or Latin squares, no simple formulae for the missing values can be given, since there are several possible patterns: the blanks in the table of results might all be in the same block, or all in different blocks, or anything between these extremes, and similarly with respect to treatments. General methods of estimation are given in more advanced texts. The fact that missing values *can* be estimated should never be made an excuse for allowing them to arise, because some precision is lost whenever there are any missing observations, since $t_{(f)}$, $\hat{\sigma}^2$ and the square root are all affected in the expression for significant differences.

In *Completely Randomized* designs, there is no estimation problem as the treatments form the only source of systematic variation. The analysis of a completely randomized experiment with unequal replication is given in Example 16.2, and this analysis applies equally whether the inequality was deliberate (as in the example) or accidental (due to lost observations).

EXAMPLE 16.2

Three culture media, A, B, C, are being used for growing colonies of fungal mycelium, and are being compared with a standard method, O, which is replicated ten times while A, B, C are replicated only five times each. After 48 hours' incubation, the radii of the growing colonies are as follows (in units of the graduations in the microscope used to examine them):

A: 25, 23, 20, 24, 27
B: 25, 28, 28, 24, 23
C: 21, 18, 15, 19, 20
O: 24, 21, 18, 20, 23, 21, 16, 19, 22, 17

We require to know whether any of A, B, C differ from the standard, O. The treatment totals are obtained as usual: $T_A = 119$, $T_B = 128$, $T_C = 93$, $T_O = 201$, so that $G = 541$. The total sum of squares of all 25 observations is also obtained in the usual way, and is $25^2 + 23^2 + 20^2 + \cdots + 19^2 + 22^2 + 17^2 - (541)^2/25 = 301\cdot76$. No alternative or shortcut method is available for calculating the treatments sum of squares in this case, and it is necessary to square each treatment total, divide by the number of observations upon which it was based, add together these contributions for all the treatments, and finally subtract G^2/N as usual. So we obtain

$$\frac{T_A^2}{5}+\frac{T_B^2}{5}+\frac{T_C^2}{5}+\frac{T_O^2}{10}-\frac{G^2}{25}$$

$$=\frac{119^2}{5}+\frac{128^2}{5}+\frac{93^2}{5}+\frac{201^2}{10}-\frac{541^2}{25}=171{\cdot}66$$

In the Analysis of Variance of Table 16.2, the d.f. for treatments and total are 3 and 24, since there are four treatments and twenty-five observations in all; by subtraction, error has 21 d.f.

Table 16.2   Analysis of Variance for the data of Example 16.2

Source of variation	D.F.	S.S.	M.S.	
Treatments (Media)	3	171·66	57·22	$F_{(3,\,21)} = 9{\cdot}23$***
Error	21	130·10	6·20 (estimate of σ^2)	
Total	24	301·76		

In assessing the results of this experiment, we wish to compare O (which had 10 observations) with any of A, B or C (having 5), for which the significant difference is $t_{(21)}\sqrt{\hat{\sigma}^2(\frac{1}{10}+\frac{1}{5})}=t_{(21)}\sqrt{3\hat{\sigma}^2/10}$. The values of $t_{(21)}$ are 2·080 (at the 5% level), 2·831 (1%), 3·819 (0·1%), and $\sqrt{3\hat{\sigma}^2/10}=1{\cdot}37$; significant differences therefore are 2·85 (at 5%), 3·88 (at 1%), 5·23 (at 0·1%). The treatment means are: A, 23·8; B, 25·6; C, 18·6; O, 20·1; so that A and B seem to give significantly higher growth than O (significance at the 5% and 0·1% levels respectively).

NON-NORMALITY

In Chapter 12, we listed the conditions that must be satisfied by the terms in our mathematical models if the analyses described in this book are to be valid (page 112). One condition was that the experimental-error term e_{tj} should follow a normal distribution, with a variance σ^2 that is constant over all the unit-plots or items used in the experiment. There are two ways in which this requirement may be violated: we may know that the unit-plots, if treated all alike, would produce records which follow some other common distribution, such as the binomial or Poisson, rather than the normal; or it may be clear that some of the treatments induce a more variable response than others. Examples of the latter situation are when the responses to some treatments are much larger than the responses to others, and the responses become more variable as they get larger. We have seen (page 112) how to test whether variances are constant or not, for the different treatments; the problem now is whether we can do anything to make analysis still possible without excessive complication when variances are *not* constant.

The statistical approach to this problem is to analyse, not the original observations, but some mathematical **transformation** of them, such as their square roots or their logarithms; the aim is to find a transformation to a new scale of measurement upon which *the variance will be constant* for all the experimental units. We summarize here the ways of treating some common cases. When data follow the *Poisson* distribution, use the square roots of the observed records, i.e. $\sqrt{y_{ij}}$ instead of y_{ij} in the analysis; for the *binomial*, the *angular transformation* of Fisher and Yates'[7] Table X is appropriate; this should therefore be applied to data expressed as proportions or percentages. If it seems that variability increases proportionately to the size of response, there are two common cases: should the *variance* seem to increase at the same rate as the size of response, the square root is used, for equality of mean and variance is a property of the Poisson distribution; or if the *standard deviation* increases at the same rate as the size, log y is appropriate. This logarithmic transformation often applies to counts of numbers of insects per leaf or per plant; to remove the difficulty of dealing with log 0 (which is not defined), log $(y+1)$ may be used—that is, 1 is added to all y-values before taking logarithms.

When transformations have been made, all analyses and significance tests must be carried out on the new (transformed) scale of measurement. This will give statistically sound answers, though unfortunately these new scales are not easy to appreciate: having made significance tests to answer the questions set or test the hypotheses laid down, it is best to rely on verbal explanation of the results, and if tables of means, etc. are needed, a qualified statistician should be asked to help prepare them. There is nothing magic about transformations: their purpose is simply to make the basic models used, such as (1)–(4) of Chapters 12, 13, valid for as many types of data as possible, thus applying statistical methods developed for the normal distribution and not having to redevelop methods for numerous other distributions.

EXERCISES

(For Answers and Comments, see p. 174)

16.1 In the data of Exercise 13.2 (p. 123), suppose that the plant receiving compound A in block V died before the experiment was completed (for a reason not due to treatment), but the remaining 29 observations were exactly as given in that Exercise. Estimate a missing value for this entry in the table, and explain how the rest of the analysis would be carried out.

16.2 In the data of Exercise 15.2 (p. 140), suppose that the entry 13 in block V were lost. Estimate a missing value to replace it, and state what effect this would have on the rest of the analysis.

16.3 The concentrations of a substance needed to produce a certain reaction in guinea-pigs, which had previously received one of the four treatments A–D, were as follows:

A: 23, 21, 26, 23 units
B: 19, 24, 22, 26, 21, 27 units
C: 25, 28, 28, 24 units
D: 19, 21, 24, 20, 22 units

Test whether there is any difference due to pre-treatment.

How would a test for the difference between the means of C and D be carried out?

16.4 In a completely randomized experiment, three treatments A, B, C are used, each in 8 replicates, and a measurement is made which takes the following values.

A: 12,   8, 15, 10, 17, 19, 11, 16
B: 32, 20, 28, 37, 25, 22, 29, 34
C: 49, 60, 59, 44, 40, 52, 46, 48

Do these data require a transformation before they can be used in an analysis of variance? If so, which one?

16.5 Apply Bartlett's test (page 112) to the data used in Example 12.1 (page 105), and comment on the result.

17

Nonparametric Significance Tests

With one exception, all the significance tests described so far have depended on some assumptions about the distribution from which the sample of observations has been drawn. The exception was the χ^2 test when used either to test the goodness-of-fit of data to a ratio-type hypothesis (Test 8.7) or to examine data arranged in a contingency table (Test 8.8). Generally we have required that observations shall be taken from a normal distribution, or at least from one which approximates closely to normality.

When there is doubt whether the basic assumptions for the normal deviate, t or F tests apply, or the assumptions for χ^2 Test 8.4, it may be appropriate to consider tests which do not depend on these assumptions, or indeed on any assumption at all about the form of distribution from which observations have been taken. However, in view of the Central Limit Theorem, the tests given for *sample means* in Chapters 7 and 8 apply very widely. The t-test can be used, even for quite small samples, when data have been drawn from a distribution that may not be normal, provided that the distribution is not too skew; under these conditions, a test on the sample mean made using t leads to very nearly the same result as an exact one worked out specially for that distribution. We may call the t-test *robust* to departures from normality; the general definition of a robust test is one which continues to give results correct at almost the same levels of significance even when some of its basic assumptions do not hold. For a symmetrical, or nearly symmetrical, distribution we may safely use t, and so it is really only for rather skew sets of data that we should look seriously for alternative tests. Even skew data can sometimes be transformed to symmetry, as we saw in Chapter 6 where the lognormal distribution was described. In such a case, a t-test could still be used on the logarithms of the original observations.

However, tests of the *shape* of a distribution (variance and skewness) are much less *robust*, and much more sensitive to non-normality. Furthermore, if we wish to compare two samples from distributions which may be of different shape, it is useful to have a test of the difference in their locations (*centres*) which does not take shape into account at all, even to assume any similarity.

The mean and variance of a distribution are, as we have seen, always closely related to the parameters in its probability density function; tests not employing these directly are called **nonparametric**, and they generally depend on **ranking**, that is arranging a set of observations in order of size, rather than using their actual numerical values. Some of the available nonparametric tests are of high efficiency, in the sense that when the assumptions for the corresponding parametric test (e.g. the t-test when means are being examined) *are* satisfied we would not lose very much in sensitivity by applying the nonparametric one instead. So if we have any doubt about making these assumptions, we may prefer to use a nonparametric test anyway, especially if it is easy and quick to apply.

LOCATION OF THE CENTRE OF A DISTRIBUTION

Corresponding to the t-Test 8.1, there is a test of whether a given sample of data came from a distribution with given median M. We use M rather than the mean, because M is more likely to be a stable measure of central tendency when the shape of a distribution is in doubt.

Test 17.1 (*The Sign Test*) A set of observations $x_1, x_2, \ldots, x_N$ is given. Test the N.H. that they arise from a distribution whose median is M. Attach a sign to each one of $x_1, x_2 \ldots, x_N$, $+$ if its numerical value is greater than M and $-$ if it is less than M. (Any x_i which is exactly equal to M must be ignored.) Denote the number of $+$ signs by n_+, and suppose that n signs have been allocated altogether ($n = N$ if no x_i was equal to M). Now if the true value of the median is M, and we have ignored any x_i that is equal to M, the probability that any x_i will be less than M is the same as the probability that it will be greater than M, namely $\frac{1}{2}$. On this hypothesis, therefore, n_+ follows a binomial distribution whose parameters are n and $\frac{1}{2}$, and we may determine whether the observed value of n_+ lies in the extreme tail of this binomial distribution.

EXAMPLE 17.1

Test the N.H. that the median of the following 21 observations has the value 10.

13, 12, 11, 9, 12, 8, 13, 12, 11, 11, 12,

10, 13, 11, 10, 14, 10, 10, 9, 11, 11

We attach signs as follows to these figures:

+, +, +, −, +, −, +, +, +, +, +,

·, +, +, ·, +, ·, ·, −, +, +

So $n = 17$ signs have been allocated, of which $n_+ = 14$ were positive. We have to discover whether the value $n_+ = 14$ lies in the tail area of the binomial distribution with $n = 17$ and $p = \frac{1}{2}$. We have no specific Alternative Hypothesis, and so we must assume that we could have an excess of minus signs just as well as of plus signs; a two-tailed significance test will therefore be needed. Remembering that the binomial distribution with $p = \frac{1}{2}$ is symmetrical, the top tail from 14 upwards and the bottom tail from 3 downwards contain all the possible values of n_+ which are as extreme as our observation or more so. Thus if $P(0) + P(1) + P(2) + P(3) + P(14) + P(15) + P(16) + P(17)$ is less than 0·05, our result is significant at the 5% level, since it falls in the extreme 5% of the binomial having $n = 17$, $p = \frac{1}{2}$. Calculations like this are tedious, though there exist tables of the binomial distribution which minimize the labour; some of the more sophisticated pocket calculators also contain facilities for finding individual terms in a binomial distribution.

However, because $p = \frac{1}{2}$ the binomial distribution is symmetrical, and so for relatively small n it is possible to use a normal approximation. In this example, $n = 17$, $p = \frac{1}{2}$, so $np = 8·5$ and $npq = 4·25$. Therefore n_+ is distributed approximately as $\mathcal{N}(8·5, 4·25)$, so we must discover whether 14 (the observed value of n_+) could reasonably arise from this distribution. Thus $d = (14·0 − 8·5)/\sqrt{4·25} = 5·5/2·06 = 2·67$ should be $\mathcal{N}(0, 1)$; this value is significant at the 1% level, suggesting that the N.H. is wrong and M is not 10.

If an Alternative Hypothesis was proposed that demanded a one-tailed test, we should proceed just as described in Chapter 7 (pages 54–55).

TWO SAMPLES, 'PAIRED'

In some experiments, data come naturally in related pairs, rather than independent items. For example, in an experiment in which a (non-absorbent) fungicide is sprayed on to the leaves of tomato plants we may choose N leaves at random from a collection of leaves of similar age on plants growing in a greenhouse; then one surface of each leaf will be coated with fungicide and the other surface not. If we can actually count the number of developing fungal spots or lesions after a fixed period of growth in an environment infested with fungal spores, we have two parametric analyses open to us. We may carry out a randomized block analysis for two treatments (fungicide and no fungicide) in N blocks, the blocks term removing any systematic differences between the leaves. We may also calculate the difference x between the number of lesions on the untreated surface and that on the treated surface for each leaf, so giving N readings of x which we analyse as in Chapter 8. Taking the difference in each leaf removes between-leaf differences: we do *not* have two independent samples, each of N observations, because they arise from the *same* leaves.

So as to make these analyses, we need to assume that the lesion counts are normally distributed, or at least that the differences x are; sometimes a logarithmic or square-root transformation (Chapter 16) will help us to justify doing the analysis. But if we are not able to justify any of these methods, or if it is very difficult to count lesions, we may choose a simpler nonparametric method, illustrated by the following example.

EXAMPLE 17.2

Forty leaves were used in a fungicide experiment, one surface of each leaf being coated with a standard (non-absorbent) fungicide, and after a week's exposure to infection the leaves were assessed as 'coated surface cleaner', 'no difference' or 'coated surface more infected'; the numbers in these categories were 28, 8 and 4 respectively. Test whether the fungicide coating is effective.

The Sign Test 17·1 may again be used; $n = 32$, ignoring the 'no difference' leaves, and $n_+ = 28$, leading to $d = (28 - 16)/\sqrt{8} = 12/2 \cdot 828 = 4 \cdot 24$, a value significant at the $0 \cdot 1 \%$ level, leading us to reject the Null Hypothesis of no difference.

However, we do not always want to throw away *all* the information about the *sizes* of differences in pairs, as the Sign Test does, and if such information is available Wilcoxon's *Matched-Pairs Signed-Ranks* test can be used.

Test 17.2 (*Signed Ranks Test*) On each of N items, two records are available: x_1 and y_1 on item 1, x_2 and y_2 on item 2, and so on. The differences $w_i = x_i - y_i$ are found, and the *moduli* of these w_i are ranked in order, ignoring any w_i that are zero. The sum of ranks for the positive w_i is calculated; call this R_+. Likewise the sum of ranks for the negative w_i is called R_-; the smaller of R_+, R_- is called T. To test the Null Hypothesis that the median difference between x and y is zero, test T as an approximately normal variate with mean $\frac{1}{4}n(n+1)$ and variance $\frac{1}{24}n(n+1)(2n+1)$; this can be used for $n = 10$ or more, n being N minus the number of zero w's.

EXAMPLE 17.3

Two red food colouring agents are compared for retention of colour in various solutions after various heat treatments, with the following results.

Solution/Treatment combination	(1)	(2)	(3)	(4)	(5)	(6)	(7)	(8)	(9)	(10)
Agent A	4	4	5	1	3	3	5	6	6	3
Agent B	7	6	9	5	6	3	4	7	4	7

The records are scores on a visual ten-point scale of colour scoring. Test whether the two agents appear to differ.

If we denote score B by x and score A by y, the values of w are 3, 2, 4, 4, 3, 0, -1, 1, -2, 4 and the rank order for the moduli of w (ignoring $-$ signs) is $5\frac{1}{2}$, $3\frac{1}{2}$, 8, 8, $5\frac{1}{2}$, $-$, $1\frac{1}{2}$, $1\frac{1}{2}$, $3\frac{1}{2}$, 8. One w was zero, and does not feature in the ranking. Several cases of *tied ranks* occur; there are two values of $|w|$ equal to 1, and so two contestants for first place in the rank order. We compromise by letting them share the first two places, giving each a rank of $1\frac{1}{2}$. Similarly there are two contestants for the next place, so they must share third and fourth, with rank $3\frac{1}{2}$; two must share fifth and sixth, with rank $5\frac{1}{2}$; and the final three are *all* contenders for seventh place, so they must share seventh, eighth and ninth, with average rank $(7+8+9)/3 = 8$. The two negative values of w have ranks $1\frac{1}{2}$ and $3\frac{1}{2}$, so $R_- = 1\frac{1}{2} + 3\frac{1}{2} = 5$. The seven positive values of w have ranks which sum to $R_+ = 40$. Here $n = 9$, and we can check our calculation by seeing that $R_- + R_+ = \frac{1}{2}n(n+1) = \frac{1}{2} \times 9 \times 10 = 45$, which it does. T is the smaller of R_-, R_+, and so is 5; we will use the normal approximation even though n is only 9 in this example. (In fact the result that it gives is very similar to that from a more exact and laborious method.)

On the Null Hypothesis that the two agents are equally good, T is, then, approximately normally distributed with mean $\frac{1}{4}n(n+1) = \frac{1}{4} \times 9 \times 10 = 22\cdot5$ and variance $\frac{1}{24}n(n+1)(2n+1) = \frac{1}{24} \times 9 \times 10 \times 19 = 71\cdot25$, so that $d = (T - 22\cdot5)/\sqrt{71\cdot25}$ is approximately $\mathcal{N}(0, 1)$. The value of d is $(5 - 22\cdot5)/8\cdot44 = -2\cdot07$, which is significant at the 5% level, so that we reject the Null Hypothesis.

The Signed Ranks test compares the medians of two populations, which are of similar shape; however, this shape need *not* be normal. It uses information about the size of the observations as well as ranking, and as a result it is much more efficient than the sign test; in fact it is not much less accurate than a *t*-test even when the assumptions for a *t*-test are valid. Therefore it is a very useful test for 'paired' data.

TWO SAMPLES, NOT 'PAIRED': THE U TEST

Test 17.3 This compares the medians of two independent samples, which need not contain the same number of observations; the Null Hypothesis to be tested is that the two medians are equal. Call the two samples I, II. Rank their members into one single order, and count the total number (U) of times that members of II precede members of I in this ranking. When neither sample size is less than 8, this count U is distributed approximately as $\mathcal{N}$ ($\frac{1}{2}n_1n_2$, $\frac{1}{12}n_1n_2(n_1 + n_2 + 1)$), n_1 and n_2 being the two sample sizes. If n_1 and n_2 are 7 or less, a special table is needed (see, for example, the book by Siegel[8]).

EXAMPLE 17.4

Test whether the medians of the following samples I, II, each of 10 observations, are equal.

I: 17·2, 5·1, 12·3, 8·2, 13·5, 13·3, 11·6, 15·2, 10·8, 7·1
II: 19·0, 12·4, 17·5, 12·8, 13·0, 10·6, 4·7, 9·3, 16·8, 10·1

In the joint ranking which follows, each member is identified as coming from sample I or sample II:

4·7 5·1 7·1 8·2 9·3 10·1 10·6 10·8 11·6 12·3
(II) (I) (I) (I) (II) (II) (II) (I) (I) (I)
12·4 12·8 13·0 13·3 13·5 15·2 16·8 17·2 17·5 19·0
(II) (II) (II) (I) (I) (I) (II) (I) (II) (II)

We now count the number of times a II precedes a I: for the value 4·7, there are 10 I's to the right of it, so score 10; for the value 9·3, there are 7 I's to its right, so score 7; for 10·1, score 7; for 10·6, score 7; for 12·4, 12·8, 13·0, score 4 each; for 16·8, score 1; for 17·5, 19·0 score 0 each. So the total score is $U = 44$. Thus $U = 44$ should be a member of $\mathcal{N}(\frac{1}{2} \times 10 \times 10, \frac{1}{12} \times 10 \times 10 \times 21)$, i.e. of $\mathcal{N}(50, 175)$, and therefore $d = (44 - 50)/\sqrt{175} = -6\cdot0/13\cdot2$ should be $\mathcal{N}(0, 1)$. The numerical value of d is well below 1, and so certainly not significant; hence we may accept the N.H. that the two samples do have the same median (whose *value* is not actually specified).

When two of the observed values are equal, we call this a *tie*; if there should happen to be ties between observations in the same sample, the test is not affected, but if there are ties between observations in different samples, this obviously complicates the problem of deciding how many times the observations from one sample precede those from the other in the joint ranking. This complication is solved by adjusting the formula for the variance of U; we omit details and refer readers to the text by Siegel.[8]

It can be shown that when the parent distributions are normal, U provides nearly as good a test for differences between means as does t (it is, in fact, about 95% efficient). But unfortunately, the U-test is only satisfactory when we are comparing the medians of samples from two distributions which have the same shape. The value of the test lies in the fact that this shape need not be normal or even symmetrical; however it is not a satisfactory test for comparing, for example, the median of a symmetrical distribution with the median of a skew one, nor even the median of a slightly skew distribution with the median of a very skew one. In fact, U is less *robust* for changes in shape than is t for the effects of non-normality, and U cannot be improved by increasing the sample sizes—a most unusual situation in significance testing.

EXERCISES

(For Answers and Comments, see p. 175)

17.1 Twenty-five subjects undergoing a test reported the following reaction times (sec) to a stimulus:

6·6, 3·6, 5·4, 7·2, 4·7, 13·1, 2·0, 7·6, 2·3, 2·8, 15·4, 4·3, 6·7, 9·5, 11·8, 1·4, 19·7, 3·0, 7·5, 6·9, 23·3, 6·4, 14·1, 6·0, 3·8.

Test the hypothesis that the median reaction time is 7·8 sec.
Calculate the mean reaction time and comment on the result.

17.2 Two samples A, B, of plants of the same species growing on opposite slopes of a valley were dug up and weighed, as follows.

A (20 observations): 27·1, 40·3, 15·7, 3·9, 22·2, 36·4, 11·8, 16·3, 14·7, 16·2, 32·0, 15·7, 12·9, 27·5, 9·9, 14·4, 24·8, 7·2, 21·0, 18·8.
B (12 observations): 11·7, 15·3, 19·1, 22·0, 6·7, 14·1, 19·1, 24·4, 15·8, 12·3, 28·7, 17·9.

Test the hypothesis that their median weights were equal.

17.3 In a tasting experiment, people were presented with pairs of samples of vegetables preserved by the same method but kept for different periods of time. It was required to test the hypothesis that no difference could be detected between samples kept for a short period of time and those kept for longer. Thirty people took part, eight said there was no difference, seven preferred the longer-kept samples, and the remaining 15 preferred the shorter-kept samples. Test the hypothesis.

17.4 The people taking part in the experiment described in the previous

Exercise (17.3) were also asked to score each sample on a ten-point scale. They obtained these results:

Person	1	2	3	4	5	6	7	8	9	10	11	12	13	14	15
Short storage	3	6	5	7	8	5	4	2	7	6	4	5	8	6	5
Long storage	4	6	4	3	3	5	5	2	5	5	4	1	3	8	5

Person	16	17	18	19	20	21	22	23	24	25	26	27	28	29	30
Short storage	8	2	6	3	6	5	4	7	5	5	7	4	5	7	6
Long storage	2	5	5	3	7	3	7	7	6	4	4	1	5	6	2

Make a further statistical test, which uses the information contained in the scores, to examine the same hypothesis as previously.

17.5   In the experiment described in the previous two Exercises (17.3, 17.4), no alternative hypothesis was offered as to which samples ought to be 'better'. A food technologist, looking at the results, claims that if there is any difference it must be the shorter-kept samples which are preferred. Examine the results (a) using only the preferences, and (b) using the scores also, in the light of this claim.

A Guide to Choosing a Significance Test

Simple tests are based on specific hypotheses about the central tendency (mean or median) of a distribution or about its scatter (variance).

1 Central tendency of a distribution

(a) Is/are the sample(s) of observations large? If so, the sample mean is a satisfactory basis for a test. See page 51 and Test 7.3 for examining the N.H. 'mean $= \mu$'; if the variance σ^2 is not known use s^2 instead. The original distribution of x need not be known, but samples can be smaller if x has a fairly symmetrical distribution. See page 63 and Test 8.3 for examining the NH 'difference between two means is $\mu_1 - \mu_2$'.

(b) If samples are not large, is the original x distribution normal (or very close to it)? If so, is the variance σ^2 known or must it be estimated by s^2 from the sample? For a single sample from a normal distribution with known variance σ^2, see Test 7.3 for examining the N.H. 'mean $= \mu$', and for two samples from normal distributions with known variances see page 62 for a test of the N.H. 'difference between two means is $\mu_1 - \mu_2$'.
For a single sample from a normal distribution with unknown variance, see Test 8.1 for examining the N.H. 'mean $= \mu$'. For two samples from normal distributions with the *same* (unknown) variance, see page 61 and Test 8.2 for examining the N.H. 'difference in means is $\mu_1 - \mu_2$' (where $\mu_1 - \mu_2$ is often 0). For several samples from normal distributions with the same (unknown) variance, use Analysis of Variance methods (Chapters 12, 13). If variances are different, see Tests 8.2a and 8.3.

(c) If samples are not large and the original x distribution is not normal, test hypotheses based on medians rather than means, by nonparametric methods.
For a single sample, Test 17.1 examines the N.H. 'median $= M$', and for two samples Test 17.3 examines the N.H. that the two medians are equal; Test 17.2 applies when ranks only (not measurements) are available.

2 Variability in a distribution

(a) Are samples from normal distributions? If not, no simple test is available.

(b) For one sample from a normal distribution, see Test 8.4 for examining the N.H. 'variance $= \sigma^2$'. For two samples from normal distributions, see Test 8.9 for examining the N.H. 'variances in both distributions are the same'.

For several samples from normal distributions, see page 112 and Example 12.2 for a test (Bartlett's) of whether all variances are the same.

3 Tests on proportions

See page 53 for testing a hypothesis about a single proportion, and pages 68–69 for comparing two proportions.

References and Bibliography

REFERENCES

1 CAUSTON, D. R. (1977). *A Biologist's Mathematics*. Edward Arnold, London.
2 CLARKE, G. M. and COOKE, D. (1978). *A Basic Course in Statistics*. Edward Arnold, London
3 COCHRAN, W. G. and COX, G. M. (1957). *Experimental Designs*, 2nd edn. Wiley, New York and Chichester.
4 DUNCAN, D. B. (1955). *Biometrics*, *11*, 1-42.
5 EDWARDS, A. L. (1968). *Experimental Design in Psychological Research*. Holt, Rinehart and Winston, New York.
6 FISHER, R. A. (1950). *Statistical Methods for Research Workers*, 11th edn. Oliver and Boyd, Edinburgh.
7 FISHER, R. A. and YATES, F. (1963). *Statistical Tables for Biological, Agricultural and Medical Research*, 6th edn. Oliver and Boyd, Edinburgh.
8 SIEGEL, S. (1956). *Nonparametric Statistics for the Behavioural Sciences*. McGraw-Hill, New York and Maidenhead.

BIBLIOGRAPHY

For a detailed study of experimental design and analysis, the book by Cochran and Cox mentioned above should be consulted.

A classical text on the methods and applications of statistics is: SNEDECOR, G. W. and COCHRAN, W. G. (1963). *Statistical Methods*, 6th edn. Iowa State College Press.

Those whose interests are in the area of medical statistics should consult: ARMITAGE, P. (1971). *Statistical Methods in Medical Research*. Blackwell, Oxford. This is also a very useful general text.

Answers to and Comments upon Exercises

Chapter 1

1.1 (a) Discover type of higher education (university, training or technical college), full-time or part-time, subject of study, year (undergraduate or postgraduate), extent to which grant-aided; all these factors may have an effect on expenditure.

(b) Ensure same age and variety of plant, same nutrition, environments entirely similar.

(c) Similar to (b); ensure also all or none of area treated with insecticide.

(d) Leaves at same stage of growth, e.g. second from top in each selected plant; no damage by pests or diseases.

1.2 (a) Expenditure discrete, but if recorded to high degree of accuracy can be treated as continuous.

(b) Number of leaves discrete (convention needed as to how far developed a leaf should be before it is recorded as such).

(c) Number of black-fly per leaf discrete.

(d) Nitrogen content continuous.

(Example 2) Yield continuous.

1.3 There should be 20 of each digit 0–9; 0 should be followed equally often by any other digit *including* itself, and similarly for 1–9 (check by looking at 00, 11, 22, ... up to 99 together, which provide sufficient frequency for study). The last check is the usual point of breakdown of *haphazard* writing.

1.4 (a) As in text, number population 001–750 and discard random choices above 750.

(b) Number population so that 251, 501, 751 are all equivalent to 001;

each population member thus has four sets of digits attached to it, so if 967 is picked we use member number 217 (= 967 − 750), if 622 use 122 (= 622 − 500), etc.

(c) Let 301–600 and 601–900 each be equivalent to 001–300, giving 3 sets of digits for each member, but discard the *incomplete* set 901–999 (and 000) since every member must have the same number of sets associated with it: hence 967 discarded, 622 becomes·022, etc.

(d) Split field into 25 × 40 = 1000 units; number these in a suitable way; proceed as in text.

1.5 (a) (d) (e) quantitative, continuous; (b) (j) quantitative, discrete; (c) (f) (g) (h) (k) qualitative.

1.6 (a) gives random sample of people, not of households (which vary in size). Need to construct list of households by using addresses given in voters' list; then sample from this.

(b) gives random sample of families, not of children (those in large families having less chance of selection). Need to construct complete list of children first, from the cards, and then sample from this.

(c) is very time-saving and may be acceptable if field very uniform; serious risk of systematic trends through field being picked up in sample, especially if rows contain a multiple of 10 plants each. No real alternative to random sample after first numbering all plants.

1.7 Group in pairs: 07 = G; 43 minus 26 = 17 = Q; 55 minus 26 minus 26 = 03 = C; 27 minus 26 = 01 = A; 18 = R; 34 minus 26 = 08 = H; 94 cannot be used, as there are only three *complete* sets of 26 numbers (01–26, 27–52, 53–78), so to use 79 upwards would give some letters higher chance of selection than others; 56 minus 26 minus 26 = 04 = D; 23 = W. Sample is G, Q, C, A, R, H, D, W.

Chapter 2

2.1 $r = 0$ 1 2 3 4 5 (the distribution is Binomial).
$f = 3$ 13 30 33 17 4

2.2 Taking too many classes loses regularity of pattern; taking too few provides hardly any summarization of original figures.

2.3 The data follow a skew curve, whose long tail is on the right (could be log-normal).

2.4 It gives the impression that observations are scattered throughout a class-interval rather than concentrated at points.

2.5 $x = 26$–30 31–35 36–40 41–45 46–50 51–55 56–60 61–65
$F =$ 4   9   32   90   151   181   184   187

2.6 Denote number of spades in hand by r.
$r = 0$ 1 2 3 4 5 6 7 8 9 10 11 12 13
$f = 2$ 6 20 28 23 12 4 2 1 0 1 1 0 0
Total frequency is 100.
The appropriate diagram is a bar-chart such as Fig. 2.1.

Chapter 3

3.1   Mean $= \dfrac{1}{200}(31 \times 1 + 37 \times 2 + 33 \times 3 + 30 \times 4 + 35 \times 5 + 34 \times 6)$

$$= \frac{703}{200} = 3\cdot515.$$

Variance $= \dfrac{1}{200 \times 199}[200 \times (31 \times 1 + 37 \times 4 + 33 \times 9$

$$+ 30 \times 16 + 35 \times 25 + 34 \times 36) - 703^2]$$

$$= \frac{1}{39800}(200 \times 3055 - 494209) = \frac{116791}{39800} = 2\cdot934.$$

Standard deviation $= +\sqrt{2\cdot934} = 1\cdot713.$

3.2   Mean $= \dfrac{1700}{370} = 4\cdot59.$ Variance $= 4\cdot41.$

3.3   Summary table

$x =$ 65   85   95   105   115   125   140   165   210   and hence
$f =$ 3   6   13   25   24   21   18   7   3
$F =$ 3   9   22   47   71   92   110   117   120

Mean $= \dfrac{14255}{120} = 118\cdot79.$ Variance $= 645\cdot38.$ S.D. $= 25\cdot4.$

There are 47 observations with values not greater than $109\frac{1}{2}$ (assumptions as in text about accuracy of measurements); next interval 10 units wide, contains 24 observations. We want 60th observation, so add $\frac{13}{24} \times 10 = 5\cdot42$ to $109\cdot5$ to give $M = 114\cdot92$. Distribution skew, median nearer to centre (mode) than is the mean.

3.4   Mean $= \dfrac{751}{36} = 20\cdot86.$

Variance $= \dfrac{1}{36 \times 35}(36 \times 16017 - 751^2) = \dfrac{12611}{1260} = 10\cdot009.$

S.D. $= 3\cdot164$. Median $= 20\cdot5 + \frac{1}{4} \times 1 = 20\cdot75.$

Distribution symmetrical so mean $=$ median (approx.).

3.5   Mean $= \dfrac{56}{50} = 1\cdot12.$ Variance $= \dfrac{50 \times 508 - 56^2}{50 \times 49} = 9\cdot087.$

Give no easily-understood summary because distribution is exceedingly irregular.

3.6   Mean $= \dfrac{349}{100} = 3\cdot49.$

On constructing a cumulative frequency table, find that the 50th and 51st observation in order of size both have value 3, so $M = 3$. Mode is 3.

Mean is higher than median and mode in a distribution skew to the right. (Median is not always *equal* to mode, usually between mode and mean.)

$$\text{Variance} = \frac{1}{99}\left(1533 - \frac{349^2}{100}\right) = \frac{314 \cdot 99}{99} = 3 \cdot 1817.$$

Standard deviation $= 1 \cdot 78$.

3.7 (a) Lowest 14, highest 28, so range $= (28 - 14) = 14$. Only useful to help decide number of class-intervals for summary.

(b) Lowest 0, highest 15, range $= 15$. Of virtually no use in such a skew distribution.

(c) Lowest 0, highest 11, range $= 11$. Merely tells us that the maximum possible range, 13, was not realized in this set of data. Of little use in such a case, because the theoretical value of range is known already.

3.8 New total is 810, mean $= \dfrac{810}{38} = 21 \cdot 32$.

Median $= 20 \cdot 5 + \frac{2}{4} \times 1 = 21 \cdot 0$.

Mean has increased more than median: if new observations had been very much larger, mean would have increased much more, but median would not so long as only two observations added.

3.9 Whole set: mean $= 52 \cdot 33$, standard deviation $= 4 \cdot 68$. First six: mean $= 48 \cdot 83$, S.D. $= 3 \cdot 31$; second six: mean $= 55 \cdot 83$, S.D. $= 2 \cdot 79$. S.D. of whole set large; much smaller when split into breeds.

Chapter 4

4.1 (a) $\frac{9}{16}$. (b) $\frac{9}{16} + \frac{3}{16} = \frac{12}{16} = \frac{3}{4}$. (c) First aB: $\frac{3}{16}$; in $\frac{3}{16}$ of these cases, second will be Ab, so $\frac{3}{16} \times \frac{3}{16} = \frac{9}{256}$ ($= 0 \cdot 035$). (d) There are two ways of selection, (AB, ab) and (ab, AB), so probability is $2 \times \frac{9}{16} \times \frac{3}{16}$ $= \frac{9}{128}$ ($= 0 \cdot 07$). Note that 2 is the binomial coefficient $\binom{2}{1}$.

4.2 $\frac{1}{10}$ disease-resistant and $\frac{1}{5}$ of these vigorous, so proportion is $\frac{1}{10} \times \frac{1}{5} = \frac{1}{50}$ $= 0 \cdot 02$. Probability that one is not worthy of further study $= 0 \cdot 98$. So for 100, probability is $(0 \cdot 98)^{100}$ ($= q^n$) in the binomial expression, being $P(r)$ for $r = 0$, $n = 100$; this, by logs, is $0 \cdot 132$. Mean $= np = 100 \times 0 \cdot 02 = 2$.

4.3 $p = 0 \cdot 6$, $n = 8$. We need $P(6) + P(7) + P(8)$. Answer is $0 \cdot 315$.

4.4 $n = 5$, $\bar{r} = \frac{252}{120} = 2 \cdot 1$. Thus $5\hat{p} = 2 \cdot 1$, so $\hat{p} = 0 \cdot 42$.

4.5 There are $\binom{6}{2} = \dfrac{6!}{2!4!} = \dfrac{6 \times 5}{2 \times 1} = 15$ possible pairs.

Means are a, b: $21 \cdot 5$; a, c: $28 \cdot 5$; a, d: $26 \cdot 5$; a, e: $24 \cdot 5$; a, f: $23 \cdot 0$;

b, c: $25 \cdot 0$; b, d: $23 \cdot 0$; b, e: $21 \cdot 0$; b, f: $19 \cdot 5$; c, d: $30 \cdot 0$; c, e: $28 \cdot 0$; c, f: $26 \cdot 5$; d, e: $26 \cdot 0$; d, f: $24 \cdot 5$; e, f: $22 \cdot 5$.

Mean of these is $\dfrac{370 \cdot 0}{15} = 24 \cdot 67$. Mean of six plants is $\dfrac{148}{6} = 24 \cdot 67$.

Plant a occurs in 5 of the 15 samples, so the probability that a pair picked at random will contain a is $\dfrac{5}{15} = \dfrac{1}{3}$.

4.6 (a) Binomial, because p is the same for each (independent) animal selected. $p = \frac{1}{3}$, $n = 12$.
(b) Not binomial. n is the random variate here, not a fixed sample size.
(c) Not binomial. p alters after each selection.
(d) Not binomial. p of success not constant, but should increase from run to run and session to session.
(e) Binomial with $n = 4$ and $p = \frac{1}{2}$.

4.7 $n = 5$, $p = 0 \cdot 8$. Probabilities are $P(5) = 0 \cdot 3277$, $P(4) - 0 \cdot 4096$, $P(3) = 0 \cdot 2048$, $P(2) = 0 \cdot 0512$, $P(1) = 0 \cdot 0064$, $P(0) = 0 \cdot 0003$. Prob. of 5 satisfactory $= 0 \cdot 328$ (accept); prob. of 2 or less satisfactory $= P(2) + P(1) + P(0) = 0 \cdot 058$ (reject); prob. of 4 or 3 satisfactory $= P(4) + P(3) = 0 \cdot 614$ (take further sample).

For 9 satisfactory, must have 4 in first and 5 in second, $P(4) \times P(5) = 0 \cdot 1342$. (4 in second, 5 in first not possible, as then second would not be necessary.)

Assume p does not change during whole period of sampling, and all items can be sampled independently of all others.

Chapter 5

5.1 (a) $P(0) = e^{-1} = \dfrac{1}{e} = \dfrac{1}{2 \cdot 7183} = 0 \cdot 3679$.   (b) $P(1) = \dfrac{e^{-1} 1}{1} = 0 \cdot 3679$.
(c) *Either* no particles, *or* one particle, or $\geqslant 2$ *must* be emitted, i.e. probability of one or other of these events is 1, $\therefore P(r \geqslant 2) = 1 - P(0) - P(1) = 0 \cdot 2642$.

5.2 Mean $= 0 \cdot 7$. Variance $= 3 \cdot 48$. Variance not equal to mean; in fact very much greater, so not Poisson. Nearer to Poisson if the three values $r = 5$, 8, 15 absent.

5.3 Mean $= 2 \cdot 1$. Variance $= 2 \cdot 475$. Sufficiently close on inspection to accept Poisson.

5.4 $\lambda = 1$. We require $P(0) + P(1) = 2 \times 0 \cdot 3679 = 0 \cdot 736$ approx. (as in **5.1**). Second part is binomial with $n = 10$, $p = 0 \cdot 736$, and $P(r = 5)$ is required. This $\dfrac{10!}{5!5!} (0 \cdot 736)^5 (0 \cdot 264)^5$ which (by logs) reduces to approximately $0 \cdot 070$.

5.5 $P(0) = e^{-\mu}$, so $e^{-\hat{\mu}} = 0 \cdot 2$; $-\hat{\mu} = \log_e 0 \cdot 2$, $\hat{\mu} = 1 \cdot 61$.

Chapter 6

6.1 (a) $d = \dfrac{5\cdot 00 - 3\cdot 95}{\sqrt{2\cdot 25}} = \dfrac{5\cdot 00 - 3\cdot 95}{1\cdot 50} = 0\cdot 70.$

(b) $d = \dfrac{0\cdot 29 - 0\cdot 50}{0\cdot 8} = -\dfrac{0\cdot 21}{0\cdot 8} = -0\cdot 26.$

(c) $d = \dfrac{-0\cdot 47 - 1\cdot 38}{1\cdot 1} = \dfrac{-1\cdot 85}{1\cdot 1} = -1\cdot 68.$

(d) $d = \dfrac{-6\cdot 89 - (-6\cdot 50)}{0\cdot 2} = \dfrac{-0\cdot 39}{0\cdot 2} = -1\cdot 95.$

6.2 (a) Second wider (more scattered). (b) Second situated further to right on graph. (c) Second further to left on graph. (d) Second further to right and also narrower.

6.3 More nearly symmetrical than the plot of $f(x)$ against x.

6.4 (a) No: n much too small; (b) Yes: n fairly large, p near $\frac{1}{2}$; $\mathcal{N}(np, npq)$, i.e. $\mathcal{N}(48, 19\cdot 2)$; (c) Yes: n moderately large, $p=\frac{1}{2}$; $\mathcal{N}(25, 12\cdot 5)$; (d) No: p too near 0, larger value of n needed; (e) Yes: p near 1 but n large; $\mathcal{N}(270, 27)$; (f) No: p extremely small, would need n very large indeed.

Note: in cases like (a), (d), (f) a useful working rule is that a normal approximation is satisfactory if $np > 5$; so (a), (f) definitely unsuitable while (d) might *just* be satisfactory if approximated by $\mathcal{N}(5, 4\cdot 5)$.

6.5 $\bar{x}$ will be $\mathcal{N}(10, \frac{25}{64})$ i.e. $\mathcal{N}(10, 0\cdot 39)$.

6.6 $\bar{x}$ will be $\mathcal{N}(5, \frac{8}{200})$ i.e. $\mathcal{N}(5, 0\cdot 04)$ even if the original distribution is *not* normal, because sample size is large. Nothing can be said about 20 observations unless they are drawn from a normal distribution.

6.7 Average no. of particles per 5-min. period $= 10 = \lambda$; r is Poisson with mean λ, and since $\lambda > 5$ the normal $\mathcal{N}(\lambda, \lambda)$, i.e. $\mathcal{N}(10, 10)$, is a good approximation.

6.8 Bags $\mathcal{N}(5\cdot 1, 0\cdot 05^2) = \mathcal{N}(5\cdot 1, 0\cdot 0025)$. Purchase consists of $n = 10$ bags; dist. of mean bag weight per purchase is $\mathcal{N}(5\cdot 1, 0\cdot 00025)$.

6.9 $\sigma^2 = 0\cdot 5^2 = 0\cdot 25.$ We require $\sigma/\sqrt{N} = 0\cdot 1$, i.e. $\sigma^2/N = 0\cdot 01.$ Hence $N = 25.$

6.10 Mean is $10 \times 0\cdot 4536 = 4\cdot 536$ kg. S.D. also multiplied by $0\cdot 4536$, so variance by $(0\cdot 4536)^2$ to give $14\cdot 65 \times (0\cdot 4536)^2 = 3\cdot 0143$ kg^2.

Chapter 7

7.1 (a) $d = 0\cdot 70$, which is much less than $1\cdot 96$, so hypothesis acceptable; (b) $d = -0\cdot 26$ so hypothesis acceptable; (c) $d = -1\cdot 68$, accept hypothesis;   (d) $d = -1\cdot95$, on borderline of significance at 5%; remarks on p. 49 apply. (Tests 7.1, 7.2).

7.2 Mean should be $\mathcal{N}(10, \frac{25}{64})$ if N.H. true; Test 7.3 gives

$$\frac{11\cdot1 - 10}{\sqrt{\frac{25}{64}}} = \frac{1\cdot1}{5} \times 8 = 1\cdot76,$$

not significant, so accept N.H. that mean $= 10$.

7.3 On N.H. that weight increase should have mean $29\cdot8$, the mean of the 20 observations should be $\mathcal{N}(29\cdot8, \frac{25}{20}) = \mathcal{N}(29\cdot8, 1\cdot25)$. Test 7.3 gives

$$\frac{28\cdot0 - 29\cdot8}{\sqrt{1\cdot25}} = -\frac{1\cdot8}{1\cdot12} = -1\cdot61, \text{ not significant, so we can assume no}$$

difference from large population.

7.4 Ref. to Exercise 6.6. $\bar{x} = 4\cdot77$ should be $\mathcal{N}(5, 0\cdot04)$, so by Test 7.3 we have

$$\frac{4\cdot77 - 5\cdot0}{\sqrt{0\cdot04}} = -\frac{0\cdot23}{0\cdot2} = -1\cdot15, \text{ not significant, so hypothesis acceptable.}$$

Cannot be done for sample of 20 unless it came from a normal distribution.

7.5 Mean should be $\mathcal{N}(2\cdot0, 2\cdot0/50)$ i.e. $\mathcal{N}(2\cdot0, 0\cdot04)$. Test 7.3 gives

$$\frac{2\cdot2 - 2\cdot0}{\sqrt{0\cdot04}} = \frac{0\cdot2}{0\cdot2} = 1\cdot0, \text{ so hypothesis acceptable.}$$

7.6 Poisson with mean > 5 approximately normal. So Test 7.2 applies; test whether $x = 11$ is a member of $\mathcal{N}(6\cdot25, 6\cdot25)$:

$$d = \frac{11 - 6\cdot25}{\sqrt{6\cdot25}} = \frac{4\cdot75}{2\cdot5} = 1\cdot90, \text{ not quite significant, so assumption reasonable.}$$

7.7 For large samples, proportion is $\mathcal{N}(p, pq/N) = \mathcal{N}(\frac{1}{2}, 1/4N)$, where N is total no. of chromosomes examined, i.e. 600. $\mathcal{N}(\frac{1}{2}, \frac{1}{2400})$ $= \mathcal{N}(\frac{1}{2}, 0\cdot000417)$, i.e. s.d. $= 0\cdot0204$.

$\hat{p} = 0\cdot42$, so $\dfrac{\hat{p} - p}{\text{s.e.}(p)} = \dfrac{0\cdot08}{0\cdot0204} = 3\cdot92$, significant at $0\cdot1\%$. So proportion could *not* reasonably be $\frac{1}{2}$.

7.8 Mean of 10 observations should be $\mathcal{N}(110, 84/10)$, so test 114 as a member of $\mathcal{N}(110, 8\cdot4)$.

$$d = \frac{114 - 110}{\sqrt{8\cdot4}} = \frac{4}{2\cdot9} = 1\cdot38, \text{ not significant, so soil conditioner has no}$$

effect. (Note that we should do a one-tailed test because hypothesis says soil conditioner effect will be *positive*.)

7.9 Normal approximation to binomial valid for $n = 200$, $p = 0\cdot8$.

$$\hat{p} = \frac{170}{200} = 0\cdot85. \quad \sqrt{\frac{pq}{n}} = \sqrt{\frac{0\cdot8 \times 0\cdot2}{200}} = 0\cdot0283.$$

So $d = \dfrac{0\cdot 85 - 0\cdot 80}{0\cdot 0283} = 1\cdot 77$, which *is* significant at 5% since a one-tailed test is needed. Reject N.H., of no difference of new drug from old, in favour of A.H. that new drug is better.

7.10 (a) $\hat{p} = \dfrac{152}{200} = 0\cdot 76$ leads to $d = -1\cdot 41$, which implies that the N.H. can be accepted; however, negative d values are not satisfactorily explained on A.H., so must involve acceping N.H. in any case.

(b) Is an example of this, with $\hat{p} = \dfrac{140}{200} = 0\cdot 7$, leading to $d = -3\cdot 53$; but this is even more inconsistent with A.H. of improvement than with N.H. of no change.

7.11 By Test $7\cdot 3$, $d = \dfrac{99 - 102}{\sqrt{49/100}} = \dfrac{-3\cdot 0}{0\cdot 7} = -4\cdot 29$, significant at $0\cdot 1\%$ in two-tailed test, so reject N.H. of no difference. If only 5 available, *cannot* apply test, since individual weights will not be normally distributed (weights usually skew to the right). The sample of 100 was large enough to assume normality of its mean.

Chapter 8

8.1 Sample mean $= +0\cdot 4$, variance $= 0\cdot 6229$. Test is

$t_{(7)} = \dfrac{0\cdot 4 - 0\cdot 1}{\sqrt{0\cdot 6229/8}} = \dfrac{0\cdot 3}{\sqrt{0\cdot 0779}} = \dfrac{0\cdot 3}{0\cdot 28} = 1\cdot 07$, not significant, so accept hypothesis.

8.2 Given sample mean $= 5\cdot 85$, variance $4\cdot 84$. On N.H. mean is distributed as $\mathcal{N}(4, 4\cdot 84/25)$, and

$t_{(24)} = \dfrac{5\cdot 85 - 4\cdot 00}{\sqrt{4\cdot 84/25}} = \dfrac{1\cdot 85}{0\cdot 44} = 4\cdot 20$, significant at $0\cdot 1\%$, so reject N.H.

8.3 Sum of squares for $A = (N-1)s^2 = 10 \times 15\cdot 2824$; for $B = 15 \times 8\cdot 0275$; so pooled variance $= \dfrac{10 \times 15\cdot 2824 + 15 \times 8\cdot 0275}{10 + 15} = \dfrac{273\cdot 2365}{25} = 10\cdot 9295$.

Test 8.2 gives

$\dfrac{6\cdot 65 - 4\cdot 28}{\sqrt{(10\cdot 9295)(\frac{1}{11} + \frac{1}{16})}} = \dfrac{2\cdot 37}{\sqrt{1\cdot 6766}} = \dfrac{2\cdot 37}{1\cdot 29} = 1\cdot 84$ which is $t_{(25)}$, and is not significant, so accept N.H. that means were equal.

8.4 Apply Test 8.1, with hypothesis that $\mu = 0$. Mean $= 29\cdot 2/11 = 2\cdot 65$. Variance $= 15\cdot 8307$.

$\dfrac{2\cdot 65 - 0}{\sqrt{15\cdot 8307/11}} = \dfrac{2\cdot 65}{\sqrt{1\cdot 439}} = \dfrac{2\cdot 65}{1\cdot 20} = 2\cdot 21$, which is $t_{(10)}$, and is not quite significant at 5% so we cannot at once accept or reject N.H. of no difference and require more data.

8.5 Apply Test 8.4. $\sum(x_i - \bar{x})^2 = 18^2 + 21^2 + 12^2 + 16^2 + 25^2 + 20^2 - (112)^2/6$
$= 2190 - 2090\cdot67 = 99\cdot33$. $(N-1)s^2/\sigma^2 = \sum(x_i - \bar{x})^2/\sigma^2 = 99\cdot33/15 = 6\cdot62$.
χ^2 has 5 d.f., two-tailed test, upper and lower $2\frac{1}{2}\%$ points used, and χ^2 certainly not significant.

8.6 One-tailed test: upper 5% point of $\chi^2_{(5)}$ used: not significant.

8.7 $(N-1)s^2/\bar{x} = 99 \times 2\cdot475/2\cdot1 = 116\cdot68$.
$$\sqrt{2\chi^2_{(99)}} = \sqrt{2 \times 116\cdot68} = \sqrt{233\cdot36} = 15\cdot28.$$
This should be $\mathcal{N}(\sqrt{2f-1}, 1)$, i.e. $\mathcal{N}(\sqrt{197}, 1)$ or $\mathcal{N}(14\cdot04, 1)$.
$(15\cdot28 - 14\cdot04)/1 = 1\cdot24$ not significant. Accept Poisson.

8.8 $s^2 = \dfrac{1}{119}\left(1 \times 31 + 4 \times 42 + 9 \times 29 + 16 \times 10 + 25 \times 2 - \dfrac{(252)^2}{120}\right)$
$= \dfrac{670 - 529\cdot2}{119}$
$= 140\cdot8/119 = 1\cdot1832$. (Really only need sum of squares $= 140\cdot8$).
$n\hat{p}\hat{q} = 5 \times 0\cdot42 \times 0\cdot58 = 1\cdot218$. $(N-1)s^2/\sigma^2 = 140\cdot8/1\cdot218 = 115\cdot60$, again not significant as $\chi^2_{(119)}$. Accept binomial.

8.9 Table is

			Total
A	$a = 172$	$b = 78$	250
B	$c = 158$	$d = 42$	200
	330	120	450

$\chi^2_{(1)} = \dfrac{450(172 \times 42 - 158 \times 78)^2}{250 \times 200 \times 330 \times 120} = \dfrac{450 \times 5100 \times 5100}{250 \times 200 \times 330 \times 120} = \dfrac{2601}{440} = 5\cdot91$,
significant at 5%, so there *is* evidence of difference.

8.10 Table is

			Total
A	$a = 17$	$b = 8$	25
B	$c = 16$	$d = 4$	20
	33	12	45

$\chi^2_{(1)} = \dfrac{45 \times (4 \times 17 - 8 \times 16)^2}{25 \times 20 \times 33 \times 12} = \dfrac{45 \times 3600}{25 \times 20 \times 33 \times 12} = \dfrac{9}{11} = 0\cdot82$, not significant. Should apply Yates' correction to make numerator $45 \times (60 - 45/2)^2$, but this reduces χ^2 and so result still further from significance. Precision of this test depends on number of observations available: 45 totally insufficient.

8.11

Digit	0	1	2	3	4	5	6	7	8	9	Total
Obs.	85	77	83	90	69	79	80	76	84	77	800
Exp.	80	80	80	80	80	80	80	80	80	80	800

$$\sum_{0}^{9}\frac{(O-E)^2}{E}=\frac{(85-80)^2}{80}+\cdots+\frac{(77-80)^2}{80}=\frac{306}{80}=3\cdot825,\text{ which is not}$$
even approaching significance (the d.f. are 9), so accept hypothesis.

8.12

		a	b	c	d	Total			a	b	c	d	Total
Obs.	I	75	15	25	5	120	Exp.	I	66	24	21	9	120
	II	85	37	26	12	160		II	88	32	28	12	160
	III	60	28	19	13	120		III	66	24	21	9	120
		220	80	70	30	400			220	80	70	30	400

$66 = 120 \times 220/400$, $24 = 80 \times 120/400$, etc.

$$\chi^2_{(r-1)(c-1)}=\chi^2_{(6)}=\frac{(66-75)^2}{66}+\cdots+\frac{(9-13)^2}{9}=11\cdot35.\text{ Not significant, so}$$
accept genetic equivalence.

	a	b+c+d	Total
For strain II, Obs.	85	75	160
Exp.	90	70	160

$\chi^2_{(1)}=5^2(1/90+1/70)=25\times0\cdot0254=0\cdot635$, not significant and so accept $9:7$ hypothesis.

8.13 First sample, $\sum x_i=429$, $\sum x_i^2=24609$, $\therefore s_1^2=\frac{1}{9}(24609-(429)^2/10)$
$=689\cdot43$. Similarly $s_2^2=255\cdot06$. $F_{(9, 11)}=689\cdot43/255\cdot06=2\cdot70$, not quite significant at 5%, so accept N.H. of equal variances.

8.14 Equality of variances must be tested first before we can know which test to apply to means. Also variance gives information on effect on plants. For sample of 9, mean $=25\cdot78$ g, variance $=13\cdot4119$; for sample of 11, mean $=28\cdot24$ g, variance $=5\cdot2525$. Variance-ratio $\dfrac{13\cdot4119}{5\cdot2525}=2\cdot55$
n.s. as $F_{(8, 10)}$, so can pool estimates of variance and use t-test for means. Estimate $s^2=(8\times13\cdot4119+10\times5\cdot2525)/18=8\cdot8789$.

$$t_{(18)}=\frac{28\cdot24-25\cdot78}{\sqrt{8\cdot8789\,(\frac{1}{9}+\frac{1}{11})}}=\frac{2\cdot46}{\sqrt{1\cdot7937}}=\frac{2\cdot46}{1\cdot34}=1\cdot84.$$

Two-tailed test, as no information which soil type may be better; t not sig., so can accept N.H. saying that soil types give same mean dry weight. F-test also allows us to accept N.H. saying variances of dry weight same. Both tests valid only if it is reasonable to assume dry weights normally distributed.

Chapter 9

9.1 (a) $\bar{x}\pm t_{(7)}\sqrt{s^2/8}=0\cdot4\pm t_{(7)}\sqrt{0\cdot6229/8}=0\cdot4\pm t_{(7)}\sqrt{0\cdot07786}$
$=0\cdot4\pm t_{(7)}\times0\cdot279$. Now $t_{(7)}=2\cdot365$ at 5% and $3\cdot499$ at 1%. So 95% limits $0\cdot4\pm0\cdot66$, i.e. $-0\cdot26$ to $+1\cdot06$, and 99% limits $0\cdot4\pm0\cdot98$, i.e. $-0\cdot58$ to $+1\cdot38$.

(b) $\bar{x} = 5\cdot85$. $s^2 = 4\cdot84$, $N = 25$. $t_{(24)} = 2\cdot064$ at 5% and $2\cdot797$ at 1%. $\sqrt{s^2/N} = \sqrt{4\cdot84/25} = 0\cdot44$. So 95% limits are $5\cdot85 \pm 2\cdot064 \times 0\cdot44 = 5\cdot85 \pm 0\cdot91$, i.e. $4\cdot94$ to $6\cdot76$, and 99% limits are $5\cdot85 \pm 2\cdot797 \times 0\cdot44 = 5\cdot85 \pm 1\cdot23$, i.e. $4\cdot62$ to $7\cdot08$.

(c) Sample A: $\bar{x} = 6\cdot65$, $s^2 = 15\cdot2824$, $N = 11$, $t_{(10)} = 2\cdot228$ at 5% and $3\cdot169$ at 1%, so by similar calculations to above, 95% limits are $4\cdot02$ to $9\cdot28$ and 99% limits are $2\cdot91$ to $10\cdot39$.

Sample B: 95% limits are $2\cdot77$ to $5\cdot79$ and 99% limits are $2\cdot19$ to $6\cdot37$.

9.2 Mean difference $\bar{x} = 2\cdot65$, variance $= 15\cdot8307$. $N = 11$. Calculation as in Exercise 9.1, giving limits $-0\cdot02$ to $+5\cdot32$ at 95% and $-1\cdot15$ to $+6\cdot45$ at 99%.

9.3 (a) Preparation A: $\hat{p} = 172/250 = 0\cdot688$, $\therefore$ $\hat{q} = 0\cdot312$. $N = 250$. $\sqrt{\hat{p}\hat{q}/N} = 0\cdot0293$. $\hat{p} \pm 2\sqrt{\hat{p}\hat{q}/N} = 0\cdot688 \pm 0.059 = 0\cdot629$ to $0\cdot747$ (i.e. *percentage of cures to be expected in future lies between* $62\cdot9\%$ *and* $74\cdot7\%$).

B: $\hat{p} = 0\cdot79$, $N = 200$. Similar calculation gives p in range $0\cdot732$ to $0\cdot848$.

(b) $\hat{p} = 85/160 = 0\cdot531$. $N = 160$. Similar calculation gives p in range $0\cdot452$ to $0\cdot610$.

9.4 Present limits $\pm 0\cdot059$: we want $\pm 0\cdot03$. Thus need to halve present width of limits, i.e. divide it by factor 2 which requires multiplying sample size by $2^2 = 4$. So new sample size is 1000.

9.5 Poisson, mean 4. $N = 50$. Use normal approximation $\mathcal{N}(4, 4/50)$ for distribution of sample mean. Approx. 95% limits $4 \pm 2\sqrt{4/50} = 4 \pm 0\cdot57$, i.e. $3\cdot43$ to $4\cdot57$ cells.

9.6 First sample has $s^2 = 689\cdot43$. $\chi^2_{(9,\,0.025)} = 19\cdot023$, $\chi^2_{(9,\,0.975)} = 2\cdot700$, $(N-1)s^2 = 9s^2 = 6204\cdot9$. Lower limit is $6204\cdot9/19\cdot023 = 326\cdot18$; upper is $6204/2\cdot700 = 2298\cdot11$. Second sample has $s^2 = 255\cdot06$. $\chi^2_{(11,\,0.025)} = 21\cdot920$, $\chi^2_{(11,\,0.975)} = 3\cdot816$, $(N-1)s^2 = 11s^2 = 2805\cdot7$. Lower limit is $2805\cdot7/21\cdot920 = 128\cdot00$; upper is $2805\cdot7/3\cdot816 = 735\cdot25$.

9.7 (a) Simply $\mu \pm 1\cdot96\,\sigma$ (see Ch. 7), i.e. $100 \pm 1\cdot96\sqrt{10}$ or $93\cdot8$ to $106\cdot2$.

(b) $100 \pm 1\cdot96 \times \sqrt{10/10}$ or $98\cdot0$ to $102\cdot0$.

(c) $100 \pm 1\cdot96\sqrt{\dfrac{10}{100}}$ or $99\cdot4$ to $100\cdot6$.

9.8 (a) Not possible. (b) Calculate sample variance in usual way, and proceed as in Example 9.1 using 100 for $\bar{x}$. (c) As in (b) except that d can be used instead of t.

Chapter 10

10.1 $\sum x_i = 2142$, $\sum y_i = 1482$, $N = 15$, $\sum x_i^2 = 312562$, $\sum y_i^2 = 154818$, $\sum x_i y_i = 211007$.

$$r = \frac{15 \times 211007 - 2142 \times 1482}{\sqrt{(15 \times 312562 - 2142^2)(15 \times 154818 - 1482^2)}} = \frac{-9339}{\sqrt{100266 \times 125946}}$$

$$= \frac{-9339}{112375} = -0.08, \text{ not significant.}$$

10.2 $\sum x_i = 1050,$ $\sum y_i = 991,$ $N = 15,$ $\sum x_i^2 = 73560,$ $\sum y_i^2 = 65509,$ $\sum x_i y_i = 69384.$

$$r = \frac{15 \times 69384 - 1050 \times 991}{\sqrt{(15 \times 73560 - 1050^2)(15 \times 65509 - 991^2)}} = \frac{+210}{\sqrt{900 \times 554}} = \frac{+210}{706}$$

$$= +0.2975, \text{ not significant.}$$

10.3 Leaf area difficult and tedious to measure. Use height instead when comparing different plants for measurements influenced by leaf area. Height is a quick and easy indicator of leaf area in plants of the same variety.

10.4 Neither x nor y can possibly be considered even approximately normally distributed, so the size of the correlation coefficient does not indicate anything specific and cannot be tested for significance.

10.5 (a) The value is significant at 0.1% (36 d.f.). This indicates that rainfall decreases as distance inland increases. The statement is in stronger terms than are justified only on a significance test, but probably has truth in it, though it does not imply that *all* rainfall is in coastal strip.
(b) The population is heterogeneous (different breeds of sow), and it is known that reproduction depends to some extent on breed; so a single correlation coefficient for the whole population is not informative.

10.6

Ranks	A	B	C	D	E	F	G	H	J	K	L	M
I	8	6	3	7	1	9	10	11	4	2	12	5
II	6	4	10	12	1	9	11	3	8	2	5	7
d_i	2	2	-7	-5	0	0	-1	8	-4	0	7	-2

$N = 12.$ $\sum d_i^2 = 216.$ $r_s = 1 - \{(6 \times 216)/(12 \times 143)\}$
$= 0.2448.$

Rank correlation more suitable when data not jointly normally distributed: neither set of percentages here looks obviously normal. Also if any actual values are numerically wrong, ranking may reduce the effect of this, as in E (Method I) which is ranked 1 whether 89 or 59 is the correct figure. Ranking particularly useful where one or two observations are very different size from others, so dominating an ordinary correlation (see 10.7).

10.7

Ranks	I	II	III	IV	V	VI	VII	VIII
X	2	1	5	7	3	6	4	8
Y	4	5	1	7	3	6	2	8
d_i	-2	-4	4	0	0	0	2	0

$N = 8.$ $\sum d_i^2 = 40.$ $r_s = 1 - \{(6 \times 40)/(8 \times 63)\} = +0.524.$

$$r_s \sqrt{\frac{N-2}{1-r_s^2}} = 0{\cdot}524 \sqrt{\frac{6}{1-0{\cdot}2744}} = 0{\cdot}524 \times 2{\cdot}876 = 1{\cdot}51,$$

which does not approach significance when tested as $t_{(6)}$. The five small pairs of records (I, II, III, V, VII) are in reverse rank order, while the other three pairs are in the same order. In calculating r, the largest observations, numerically, dominate the calculation to give a highly significant value, $0{\cdot}996$; but the sizes X, Y are clearly not normally distributed so the significance of r cannot be tested. There are really two distinct populations of environments.

Chapter 11

11.1 $\sum x_i = 16{\cdot}5$, $\sum y_i = 209$, $N = 6$, $\therefore \bar{x} = 2{\cdot}75$, $\bar{y} = 34{\cdot}83$. $\sum x_i^2 = 66{\cdot}25$, $\sum y_i^2 = 9941$, $\sum x_i y_i = 810$, $\therefore \sum(x_i - \bar{x})^2 = 20{\cdot}875$, $\sum(x_i - \bar{x})(y_i - \bar{y}) = 235{\cdot}25$, $\sum(y_i - \bar{y})^2 = 2660{\cdot}83$. $\therefore b = 235{\cdot}25/20{\cdot}875 = 11{\cdot}27$, line is $(y - 34{\cdot}83) = 11{\cdot}27(x - 2{\cdot}75)$, i.e. $y = 11{\cdot}27x + 3{\cdot}84$. So for $x = 0$, $\frac{1}{2}$, 1, 2, 3, 4, 5, 6, we predict $y = 3{\cdot}84$, $9{\cdot}48$, $15{\cdot}11$, $26{\cdot}38$, $37{\cdot}65$, $48{\cdot}92$, $60{\cdot}19$, $71{\cdot}46$. The value $y = 3{\cdot}84$ at $x = 0$ may represent base-level of nutrients present in soil; *or* the relation between y and x may become a curved one when x approaches 0. S.S. for regression $= (235{\cdot}25)^2/20{\cdot}875 = 2651{\cdot}14$, S. S. for deviations $= 2660{\cdot}83 - 2651{\cdot}14 = 9{\cdot}69$, with 4 d.f., $\hat{\sigma}^2 = 9{\cdot}69/4 = 2{\cdot}4225$, $F_{(1, 4)} = 2651{\cdot}14/2{\cdot}4225 = 1094$, significant at $0{\cdot}1\%$ $\therefore$ line good fit. $\mathrm{Var}(\hat{b}) = 2{\cdot}4225/20{\cdot}875 = 0{\cdot}116$, s.e.$(\hat{b}) = 0{\cdot}34$. $\therefore 95\%$ limits to b are $\hat{b} \pm t_{(4, \, 0{\cdot}05)}$s.e.$(\hat{b}) = 11{\cdot}27 \pm 2{\cdot}776 \times 0{\cdot}34$ i.e. $10{\cdot}33$ to $12{\cdot}21$.

11.2 Graph suggests fitting $\log y = a + bx$. For $x = 0, 1, 2, \ldots, 8$, values of $\log y$ are $-0{\cdot}12$, $0{\cdot}08$, $0{\cdot}24$, $0{\cdot}40$, $0{\cdot}54$, $0{\cdot}67$, $0{\cdot}79$, $0{\cdot}92$, $1{\cdot}06$, with mean $0{\cdot}51$. $\bar{x} = 4{\cdot}0$. $\sum(x_i - \bar{x})^2 = 60$. $\sum(x_i - \bar{x})(Y_i - \bar{Y}) = 8{\cdot}61$. ($Y$ stands for $\log y$.) Thus $\hat{b} = 8{\cdot}61/60 = 0{\cdot}144$. $\therefore \log y - 0{\cdot}51 = 0{\cdot}144 \times (x - 4)$, i.e. $\log y = 0{\cdot}144x - 0{\cdot}066$.

11.3 Line is $y = 3{\cdot}849x - 0{\cdot}536$ (some suggestion of slight curve at two ends but line fits well). If 10 added to all x-values, a increases by 10 and b does not change. If all x increased by 9%, both a and b increase by 9%.

Chapter 12

12.1 Method totals: A, $9{\cdot}10$; B, $9{\cdot}25$; C, $9{\cdot}05$. Grand total $= 27{\cdot}40$. Total S.S. $= 0{\cdot}009533$. Methods S.S. $= 0{\cdot}004333$. $\therefore$ Error S.S. $= 0{\cdot}005200$. Methods M.S. (2 d.f.) $= 0{\cdot}00217$, Error M.S. (12 d.f.) $= 0{\cdot}00043$. $F_{(2, 12)} = 5{\cdot}05$, significant at 5%. Means: A, $1{\cdot}82$; B, $1{\cdot}85$; C, $1{\cdot}81$. B v. A:

$$\frac{1{\cdot}85 - 1{\cdot}82}{\sqrt{2 \times 0{\cdot}00043/5}} = \frac{0{\cdot}03}{\sqrt{0{\cdot}000173}} = \frac{0{\cdot}03}{0{\cdot}013} = 2{\cdot}31,$$ significant at 5% as $t_{(12)}$.

C v. A: not significant.

12.2 Total S.S. $= 2508{\cdot}72$. Varieties S.S. $= 1208{\cdot}47$. Error S.S. $= 1300{\cdot}25$.

Varieties M.S. (7 d.f.) $= 172 \cdot 64$, Error M.S. (24 d.f.) $= 54 \cdot 18$, $F_{(7, 24)}$ $= 3 \cdot 19$, significant at 5%. Means: A, $37 \cdot 3$; B, $35 \cdot 0$; C, $39 \cdot 0$; D, $30 \cdot 8$; E, $26 \cdot 8$; F, $34 \cdot 0$; G, $48 \cdot 0$; H, $30 \cdot 0$. Significant differences are $t_{(24)} \sqrt{2 \times 54 \cdot 18/4} = t_{(24)} \sqrt{27 \cdot 09} = t_{(24)} \times 5 \cdot 2$. Values of $t_{(24)}$ are $2 \cdot 064$ (5%), $2 \cdot 797$ (1%), $3 \cdot 745$ $(0 \cdot 1\%)$, so significant differences are $10 \cdot 7$ (5%), $14 \cdot 5$ (1%), $19 \cdot 5$ $(0 \cdot 1\%)$. Suggests G better than B (significant at 5%), E worse than C (significant at 5%), but variability high and more replicates desirable.

Chapter 13

13.1 Total S.S. $= 1 \cdot 3783$, Blocks S.S. $= 0 \cdot 5233$, Treatments (Densities) S.S. $= 0 \cdot 7083$, $\therefore$ Error S.S. $= 0 \cdot 1467$. Blocks M.S. (5 d.f.) $= 0 \cdot 1047$, Treatments M.S. (3 d.f.) $= 0 \cdot 2361$, Error M.S. (15 d.f.) $= 0 \cdot 0098$. For Blocks, $F_{(5, 15)} = 10 \cdot 68$, significant at $0 \cdot 1\%$, for treatments $F_{(3, 15)}$ $= 24 \cdot 09$, significant at $0 \cdot 1\%$. Blocks have removed a considerable amount of systematic variation. For treatment means, the significant differences are $t_{(15)} \sqrt{2\hat{\sigma}^2/6} = t \sqrt{0 \cdot 0033}$. $t_{(15)} = 2 \cdot 131$ at 5%, $2 \cdot 947$ at 1%, $4 \cdot 073$ at $0 \cdot 1\%$, and $\sqrt{0 \cdot 0033} = 0 \cdot 057$. So significant differences are $0 \cdot 12$ (5%), $0 \cdot 17$ (1%), $0 \cdot 23$ $(0 \cdot 1\%)$. Means are A: $2 \cdot 82$; B: $3 \cdot 00$; C: $3 \cdot 28$; D: $3 \cdot 13$. Steady increase, which is significant, from A to C, then drop to D.

13.2 Total S.S. $= 2055 \cdot 47$, Blocks S.S. $= 163 \cdot 13$, Compounds S.S. $= 1720 \cdot 67$, $\therefore$ Error S.S. $= 171 \cdot 67$. Blocks M.S. (4 d.f.) $= 40 \cdot 78$, Compounds M.S. (5 d.f.) $= 344 \cdot 13$, Error M.S. (20 d.f.) $= 8 \cdot 58$. For Blocks, $F_{(4, 20)} = 4 \cdot 75$, significant at 1%, for Compounds, $F_{(5, 20)} = 40 \cdot 11$, significant at $0 \cdot 1\%$. $\sqrt{2\hat{\sigma}^2/r} = \sqrt{2 \times 8 \cdot 58/5} = \sqrt{3 \cdot 432} = 1 \cdot 85$. $t_{(20)} = 2 \cdot 086$ (5%), $2 \cdot 845$ (1%), $3 \cdot 850$ $(0 \cdot 1\%)$, so significant differences are $3 \cdot 9$ (5%), $5 \cdot 3$ (1%), $7 \cdot 1$ $(0 \cdot 1\%)$. B, D, F not significantly different, others higher than these.

13.3 Taster totals a 83, b 84, c 85, d 84. Sessions 76, 84, 85 and 91. Suppliers A 94, B 78, C 78, D 86. So $G = 336$. S.S. for tasters $= (83^2 + 84^2 + 85^2 + 84^2)/4 - 336^2/16 = 0 \cdot 5$, with 3 d.f. S.S. for sessions $= (76^2 + 84^2 + 85^2 + 91^2)/4 - 336^2/16 = 28 \cdot 5$, with 3 d.f. S.S. for suppliers $= (94^2 + 78^2 + 78^2 + 86^2)/4 - 336^2/16 = 44 \cdot 0$, with 3 d.f. Total S.S. $= 21^2 + \ldots + 26^2 - 336^2/16 = 78 \cdot 0$, with 15 d.f. Hence $S_E = 5 \cdot 0$, with 6 d.f., and $\hat{\sigma}^2 = 5 \cdot 0/6 = 0 \cdot 833$. M.S. for tasters, sessions, suppliers are $0 \cdot 167$, $9 \cdot 500$, $14 \cdot 667$, and $F_{(3, 6)} < 1$ (n.s.) for tasters, $F_{(3, 6)} = 11 \cdot 40^{**}$ for sessions (perhaps due to steady increase in score from session 1 to 4), $F_{(3, 6)} = 17 \cdot 61^{**}$ for products, indicating an overall difference. Test any specific hypotheses about products in t-tests with 6 d.f.

Chapter 14

14.1 $\sigma = 2 \cdot 8$. Error d.f. will be $(r - 1)(t - 1) = 21$. $t_{(21, \, 0 \cdot 05)} = 2 \cdot 08$. $\delta = 2 \cdot 08 \times 2 \cdot 8 \sqrt{2/4} = 4 \cdot 12$. This represents a $22 \cdot 9\%$ change in mean.

14.2 When $V = 8\%$ and $\delta = 10\%$, $10 = 8t\sqrt{2/r}$, and using $t \doteq 2$ this gives $100r = 512$, so $r = 6$ is the minimum needed. Increasing r to 12, with $V = 12\%$, $\delta = 12t\sqrt{2/12} \doteq 24/\sqrt{6} = 9\cdot8\%$, about the same as before.

Chapter 15

15.1 Totals: Blocks 80, 85, 89, 96; Treatments ① 58, n 104, p 65, np 123. Blocks S.S. $= 34\cdot25$ (3 d.f.), N $= 676\cdot00$ (1 d.f.), P $= 42\cdot25$ (1 d.f.), NP $= 9\cdot00$ (1 d.f.), Total S.S. $= 829\cdot75$, Error S.S. $= 68\cdot25$ (9 d.f.) $\therefore$ Error M.S. $= 7\cdot58$. Blocks and NP not significant, N significant at $0\cdot1\%$, P at 5%. Mean $+ $N $= 28\cdot4$, $- $N $= 15\cdot4$, difference significant at $0\cdot1\%$; $+$P $= 23\cdot5$, $-$P $= 20\cdot3$, difference significant at 5%. (No need for t-tests, since F has only one d.f. and so measures the same thing.)

15.2 Blocks S.S. $= 26\cdot8$ (4 d.f.), Blocks M.S. $= 6\cdot7$, S.S. for A $= 304\cdot2$, B $= 192\cdot2$, AB $= 88\cdot2$ (each 1 d.f.), Total S.S. $= 751\cdot8$, Error S.S. $= 140\cdot4$ (12 d.f.), Error M.S. $= 11.7$. Blocks n.s., A, B, AB all significant. Mean for A alone $= 25\cdot0$, higher than all others, so AB interaction implies that B counterbalances beneficial effect of A when added together.

15.3 Analysis of Variance

Source of Variation	D.F.	S.S.	M.S.	
R	3	112·1250	37·375	
T	2	44·3333	22·167	
RT	6	89·0000	14·833	$F_{(6, 12)} = 3\cdot46*$
Treatments	11	245·4583		
Error	12	51·5000	$4\cdot2917 = \hat{\sigma}^2$	
Total	23	296·9583		

Interaction significant. Construct 2-way table of means. Within this table sig. diffs are $t_{(12)}\sqrt{2 \times 4\cdot2917/2} = 2\cdot072t = 4\cdot51$ (5%), $6\cdot33$ (1%), $8\cdot95$ ($0\cdot1\%$).

Means	T_1	T_2	T_3
R_1	12·0	15·0	17·5
R_2	18·0	21·5	15·5
R_3	18·0	20·5	19·5
R_4	17·5	19·5	25·0

Can compare T means at each R level, or R means at each T level, to look for reasons for interaction.

Chapter 16

16.1 $B'_V = 40$, $T'_A = 105$, $G' = 358$. $\therefore x = (5 \times 40 + 6 \times 105 - 358)/(5 \times 4) = 472/20 = 23\cdot6$. Take missing value as 24, put in table of results, reduce error d.f. to 19, total d.f. to 28.

16.2 Treat as randomized block. Incomplete $B'_V = 55$, $T' = 52$, $G' = 305$.
$\therefore x = 14 \cdot 8$. Take x as 15, reduce error d.f. to 11, total to 18. Factorial
part of analysis as usual.

16.3 Trts. S.S. $= 93^2/4 + 139^2/6 + 105^2/4 + 106^2/5 - 443^2/19 = 56 \cdot 98$. Total
S.S. $= 144 \cdot 11$. $\therefore$ Error S.S. $= 87 \cdot 13$. Treatments M.S. (3 d.f.) $= 18 \cdot 99$,
Error M.S. (15 d.f.) $= 5 \cdot 81$, $F_{(3, 15)} = 3 \cdot 27$ for Treatments, on borderline of significance at 5%. Try to obtain more data before accepting
or rejecting hypothesis of differences between treatments. To
compare C, D the variance of difference is $\hat{\sigma}^2(\frac{1}{4} + \frac{1}{5})$, so test
$(\bar{y}_C - \bar{y}_D)/\sqrt{\hat{\sigma}^2(\frac{1}{4} + \frac{1}{5})}$ as $t_{(15)}$.

16.4 For A, $\bar{y} = 13 \cdot 5$, $s = 3 \cdot 82$, $s^2 = 14 \cdot 5714$
For B, $\bar{y} = 28 \cdot 4$, $s = 5 \cdot 88$, $s^2 = 34 \cdot 5536$
For C, $\bar{y} = 49 \cdot 8$, $s = 6 \cdot 99$, $s^2 = 48 \cdot 7857$

So s^2 is roughly proportional to $\bar{y}$ for these sets of data, which suggests
that a square-root transformation would be useful.

16.5 For A, $\bar{y} = 20 \cdot 43$, $s = 4 \cdot 86$, $s^2 = 23 \cdot 6190$
For B, $\bar{y} = 41 \cdot 57$, $s = 6 \cdot 70$, $s^2 = 44 \cdot 9524$
For C, $\bar{y} = 31 \cdot 43$, $s = 5 \cdot 41$, $s^2 = 29 \cdot 2857$

In Bartlett's test, $k = 3$, each s_i^2 has $f_i = 6$ d.f., $s^{*2} = 32 \cdot 6190$.
$C = 1 + \frac{1}{6}(\frac{3}{6} - \frac{1}{18}) = 29/27$. $M = 2 \cdot 3026(18 \times 1 \cdot 5135 - 6 \times \{1 \cdot 3733 +$
$1 \cdot 6528 + 1 \cdot 4667\}) = 0 \cdot 6590$. $M/C = 0 \cdot 614$, n.s. as $\chi^2_{(2)}$.
But s^2 seems roughly proportional to $\bar{y}$ and so a square-root transformation might improve the assumption of constant variance. Treatments were however highly significantly different, and this is not
changed by the transformation. (Remember Bartlett's test not very
sensitive.)

Chapter 17

17.1 $n = 25$, $\therefore np = 12 \cdot 5$, $npq = 6 \cdot 25$. $n_+ = 18$. $(18 - 12 \cdot 5)/\sqrt{6 \cdot 25} = 5 \cdot 5/2 \cdot 5$
$= 2 \cdot 2$, significant as $\mathcal{N}(0, 1)$, so reject N.H. that $M = 7 \cdot 8$. Sample
mean $= 7 \cdot 804$, so a *t*-test is bound to accept N.H. that $\mu = 7 \cdot 8$. But
clearly sample did not come from a normal distribution, because it
has a few relatively very large values; hence *t* does not apply.

17.2 Joint ranking ABAABABABABAABAABAABABBBABABABAAABAAA.
$\frac{1}{2}n_1 n_2 = 120$. $\frac{1}{12}n_1 n_2(n_1 + n_2 + 1) = 660$. $U = 132$ (B preceding A).
$(132 - 120)/\sqrt{660} = 12/25 \cdot 7 < 1$, not significant so accept equality of
medians.

17.3 $N = 30$ people, but only $n = 22$ stated a preference. Sign test for $n = 22$,
$n_+ = 15$ gives $d = (15 - 11)/\sqrt{5 \cdot 5} = 4/2 \cdot 345 = 1 \cdot 71$ n.s. So we can accept
N.H. of no difference (two-tailed test).

17.4 Rank differences w_i are -1, 0, 1, 4, 5, 0, -1, 0, 2, 1, 0, 4, 5, -2, 0,
6, -3, 1, 0, -1, 2, -3, 0, -1, 1, 3, 3, 0, 1, 4. Ranks for $|w_i|$ are
(ignoring zeros) 5, 5, 18, $20\frac{1}{2}$, 5, 11, 5, 18, $20\frac{1}{2}$, 11, 22, $14\frac{1}{2}$, 5, 5, 11,
$14\frac{1}{2}$, 5, 5, $14\frac{1}{2}$, $14\frac{1}{2}$, 5, 18 (there are 9 values of w equal to $+1$ or -1,

so 9 values of $|w|$ equal to 1, and these must 'share' the first 9 places in rank order, so they all have average rank 5; and similarly for other tied ranks). $R_+ = 193$, $R_- = 60$, so $T = 60$. On N.H. T is $\mathcal{N}(126\cdot5, 948\cdot75)$, so $d = (60 - 126\cdot5)/\sqrt{948\cdot75}$ is $\mathcal{N}(0, 1)$; its value is $-66\cdot5/30\cdot80 = -2\cdot16$, significant at 5% level leading us to reject the N.H. of no difference.

17.5 (a) One-tailed test now needed, $d = 1\cdot71$ significant at 5% level. (b) Use one-tailed significance values, but level of significance does not change. Both methods lead to A.H. being accepted in favour of N.H. of no difference.

Table I   Student's *t*—distribution
Values exceeded in two-tailed test with probability *P*.

d.f.	P= 0·1	0·05	0·02	0·01	0·002	0·001
1	6·314	12·706	31·821	63·657	318·31	636·62
2	2·920	4·303	6·965	9·925	22·327	31·598
3	2·353	3·182	4·541	5·841	10·214	12·924
4	2·132	2·776	3·747	4·604	7·173	8·610
5	2·015	2·571	3·365	4·032	5·893	6·869
6	1·943	2·447	3·143	3·707	5·208	5·959
7	1·895	2·365	2·998	3·499	4·785	5·408
8	1·860	2·306	2·896	3·355	4·501	5·041
9	1·833	2·2€2	2·821	3·250	4·297	4·781
10	1·812	2·228	2·764	3·169	4·144	4·587
11	1·796	2·201	2·718	3·106	4·025	4·437
12	1·782	2·179	2·681	3·055	3·930	4·318
13	1·771	2·160	2·650	3·012	3·852	4·221
14	1·761	2·145	2·624	2·977	3·787	4·140
15	1·753	2·131	2·602	2·947	3·733	4·073
16	1·746	2·120	2·583	2·921	3·686	4·015
17	1·740	2·110	2·567	2·898	3·646	3·965
18	1·734	2·101	2·552	2·878	3·610	3·922
19	1·729	2·093	2·539	2·861	3·579	3·883
20	1·725	2·086	2·528	2·845	3·552	3·850
21	1·721	2·080	2·518	2·831	3·527	3·819
22	1·717	2·074	2·508	2·819	3·505	3·792
23	1·714	2·069	2·500	2·807	3·485	3·767
24	1·711	2·064	2·492	2·797	3·467	3·745
25	1·708	2·060	2·485	2·787	3·450	3·725
26	1·706	2·056	2·479	2·779	3·435	3·707
27	1·703	2·052	2·473	2·771	3·421	3·690
28	1·701	2·048	2·467	2·763	3·408	3·674
29	1·699	2·045	2·462	2·756	3·396	3·659
30	1·697	2·042	2·457	2·750	3·385	3·646
40	1·684	2·021	2·423	2·704	3·307	3·551
60	1·671	2·000	2·390	2·660	3·232	3·460
120	1·658	1·980	2·358	2·617	3·160	3·373
∞	1·645	1·960	2·326	2·576	3·090	3·291

The last row of the table (∞) gives values of *d*, the unit (standard) normal deviate.

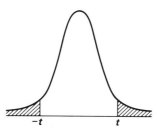

P is the sum of the two shaded areas

Table II Values of the χ^2 distribution exceeded with probability P.

P d.f.	0·995	0·975	0·050	0·025	0·010	0·005	0·001
1	392704.10⁻¹⁰	982069.10⁻⁹	3·84146	5·02389	6·63490	7·87944	10·828
2	0·0100251	0·0506356	5·99146	7·37776	9·21034	10·5966	13·816
3	0·0717218	0·215795	7·81473	9·34840	11·3449	12·8382	16·266
4	0·206989	0·484419	9·48773	11·1433	13·2767	14·8603	18·467
5	0·411742	0·831212	11·0705	12·8325	15·0863	16·7496	20·515
6	0·675727	1·23734	12·5916	14·4494	16·8119	18·5476	22·458
7	0·989256	1·68987	14·0671	16·0128	18·4753	20·2777	24·322
8	1·34441	2·17973	15·5073	17·5345	20·0902	21·9550	26·125
9	1·73493	2·70039	16·9190	19·0228	21·6660	23·5894	27·877
10	2·15586	3·24697	18·3070	20·4832	23·2093	25·1882	29·588
11	2·60322	3·81575	19·6751	21·9200	24·7250	26·7568	31·264
12	3·07382	4·40379	21·0261	23·3367	26·2170	28·2995	32·909
13	3·56503	5·00875	22·3620	24·7356	27·6882	29·8195	34·528
14	4·07467	5·62873	23·6848	26·1189	29·1412	31·3194	36·123
15	4·60092	6·26214	24·9958	27·4884	30·5779	32·8013	37·697
16	5·14221	6·90766	26·2962	28·8454	31·9999	34·2672	39·252
17	5·69722	7·56419	27·5871	30·1910	33·4087	35·7185	40·790
18	6·26480	8·23075	28·8693	31·5264	34·8053	37·1565	42·312
19	6·84397	8·90652	30·1435	32·8523	36·1909	38·5823	43·820
20	7·43384	9·59078	31·4104	34·1696	37·5662	39·9968	45·315
21	8·03365	10·28293	32·6706	35·4789	38·9322	41·4011	46·797
22	8·64272	10·9823	33·9244	36·7807	40·2894	42·7957	48·268
23	9·26043	11·6886	35·1725	38·0756	41·6384	44·1813	49·728
24	9·88623	12·4012	36·4150	39·3641	42·9798	45·5585	51·179
25	10·5197	13·1197	37·6525	40·6465	44·3141	46·9279	52·618
26	11·1602	13·8439	38·8851	41·9232	45·6417	48·2899	54·052
27	11·8076	14·5734	40·1133	43·1945	46·9629	49·6449	55·476
28	12·4613	15·3079	41·3371	44·4608	48·2782	50·9934	56·892
29	13·1211	16·0471	42·5570	45·7223	49·5879	52·3356	58·301
30	13·7867	16·7908	43·7730	46·9792	50·8922	53·6720	59·703
40	20·7065	24·4330	55·7585	59·3417	63·6907	66·7660	73·402
50	27·9907	32·3574	67·5048	71·4202	76·1539	79·4900	86·661
60	35·5345	40·4817	79·0819	83·2977	88·3794	91·9517	99·607
70	43·2752	48·7576	90·5312	95·0232	100·425	104·215	112·317
80	51·1719	57·1532	101·879	106·629	112·329	116·321	124·839
90	59·1963	65·6466	113·145	118·136	124·116	128·299	137·208
100	67·3276	74·2219	124·342	129·561	135·807	140·169	149·449

For d.f. > 100, test $\sqrt{2\chi^2_{(f)}}$ as $\mathscr{N}$ ($\sqrt{2f-1}$, 1).

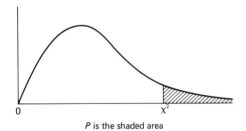

P is the shaded area

Table III   Table of *F*—distribution   Upper 5% points

ν_2 \ ν_1	1	2	3	4	5	6	7	8	9	10	12	15	20	24	30	40	60	120	∞
1	161·4	199·5	215·7	224·6	230·2	234·0	236·8	238·9	240·5	241·9	243·9	245·9	248·0	249·1	250·1	251·1	252·2	253·3	254·3
2	18·51	19·00	19·16	19·25	19·30	19·33	19·35	19·37	19·38	19·40	19·41	19·43	19·45	19·45	19·46	19·47	19·48	19·49	19·50
3	10·13	9·55	9·28	9·12	9·01	8·94	8·89	8·85	8·81	8·79	8·74	8·70	8·66	8·64	8·62	8·59	8·57	8·55	8·53
4	7·71	6·94	6·59	6·39	6·26	6·16	6·09	6·04	6·00	5·96	5·91	5·86	5·80	5·77	5·75	5·72	5·69	5·66	5·63
5	6·61	5·79	5·41	5·19	5·05	4·95	4·88	4·82	4·77	4·74	4·68	4·62	4·56	4·53	4·50	4·46	4·43	4·40	4·36
6	5·99	5·14	4·76	4·53	4·39	4·28	4·21	4·15	4·10	4·06	4·00	3·94	3·87	3·84	3·81	3·77	3·74	3·70	3·67
7	5·59	4·74	4·35	4·12	3·97	3·87	3·79	3·73	3·68	3·64	3·57	3·51	3·44	3·41	3·38	3·34	3·30	3·27	3·23
8	5·32	4·46	4·07	3·84	3·69	3·58	3·50	3·44	3·39	3·35	3·28	3·22	3·15	3·12	3·08	3·04	3·01	2·97	2·93
9	5·12	4·26	3·86	3·63	3·48	3·37	3·29	3·23	3·18	3·14	3·07	3·01	2·94	2·90	2·86	2·83	2·79	2·75	2·71
10	4·96	4·10	3·71	3·48	3·33	3·22	3·14	3·07	3·02	2·98	2·91	2·85	2·77	2·74	2·70	2·66	2·62	2·58	2·54
11	4·84	3·98	3·59	3·36	3·20	3·09	3·01	2·95	2·90	2·85	2·79	2·72	2·65	2·61	2·57	2·53	2·49	2·45	2·40
12	4·75	3·89	3·49	3·26	3·11	3·00	2·91	2·85	2·80	2·75	2·69	2·62	2·54	2·51	2·47	2·43	2·38	2·34	2·30
13	4·67	3·81	3·41	3·18	3·03	2·92	2·83	2·77	2·71	2·67	2·60	2·53	2·46	2·42	2·38	2·34	2·30	2·25	2·21
14	4·60	3·74	3·34	3·11	2·96	2·85	2·76	2·70	2·65	2·60	2·53	2·46	2·39	2·35	2·31	2·27	2·22	2·18	2·13
15	4·54	3·68	3·29	3·06	2·90	2·79	2·71	2·64	2·59	2·54	2·48	2·40	2·33	2·29	2·25	2·20	2·16	2·11	2·07
16	4·49	3·63	3·24	3·01	2·85	2·74	2·66	2·59	2·54	2·49	2·42	2·35	2·28	2·24	2·19	2·15	2·11	2·06	2·01
17	4·45	3·59	3·20	2·96	2·81	2·70	2·61	2·55	2·49	2·45	2·38	2·31	2·23	2·19	2·15	2·10	2·06	2·01	1·96
18	4·41	3·55	3·16	2·93	2·77	2·66	2·58	2·51	2·46	2·41	2·34	2·27	2·19	2·15	2·11	2·06	2·02	1·97	1·92
19	4·38	3·52	3·13	2·90	2·74	2·63	2·54	2·48	2·42	2·38	2·31	2·23	2·16	2·11	2·07	2·03	1·98	1·93	1·88
20	4·35	3·49	3·10	2·87	2·71	2·60	2·51	2·45	2·39	2·35	2·28	2·20	2·12	2·08	2·04	1·99	1·95	1·90	1·84
21	4·32	3·47	3·07	2·84	2·68	2·57	2·49	2·42	2·37	2·32	2·25	2·18	2·10	2·05	2·01	1·96	1·92	1·87	1·81
22	4·30	3·44	3·05	2·82	2·66	2·55	2·46	2·40	2·34	2·30	2·23	2·15	2·07	2·03	1·98	1·94	1·89	1·84	1·78
23	4·28	3·42	3·03	2·80	2·64	2·53	2·44	2·37	2·32	2·27	2·20	2·13	2·05	2·01	1·96	1·91	1·86	1·81	1·76
24	4·26	3·40	3·01	2·78	2·62	2·51	2·42	2·36	2·30	2·25	2·18	2·11	2·03	1·98	1·94	1·89	1·84	1·79	1·73
25	4·24	3·39	2·99	2·76	2·60	2·49	2·40	2·34	2·28	2·24	2·16	2·09	2·01	1·96	1·92	1·87	1·82	1·77	1·71
26	4·23	3·37	2·98	2·74	2·59	2·47	2·39	2·32	2·27	2·22	2·15	2·07	1·99	1·95	1·90	1·85	1·80	1·75	1·69
27	4·21	3·35	2·96	2·73	2·57	2·46	2·37	2·31	2·25	2·20	2·13	2·06	1·97	1·93	1·88	1·84	1·79	1·73	1·67
28	4·20	3·34	2·95	2·71	2·56	2·45	2·36	2·29	2·24	2·19	2·12	2·04	1·96	1·91	1·87	1·82	1·77	1·71	1·65
29	4·18	3·33	2·93	2·70	2·55	2·43	2·35	2·28	2·22	2·18	2·10	2·03	1·94	1·90	1·85	1·81	1·75	1·70	1·64
30	4·17	3·32	2·92	2·69	2·53	2·42	2·33	2·27	2·21	2·16	2·09	2·01	1·93	1·89	1·84	1·79	1·74	1·68	1·62
40	4·08	3·23	2·84	2·61	2·45	2·34	2·25	2·18	2·12	2·08	2·00	1·92	1·84	1·79	1·74	1·69	1·64	1·58	1·51
60	4·00	3·15	2·76	2·53	2·37	2·25	2·17	2·10	2·04	1·99	1·92	1·84	1·75	1·70	1·65	1·59	1·53	1·47	1·39
120	3·92	3·07	2·68	2·45	2·29	2·17	2·09	2·02	1·96	1·91	1·83	1·75	1·66	1·61	1·55	1·50	1·43	1·35	1·25
∞	3·84	3·00	2·60	2·37	2·21	2·10	2·01	1·94	1·88	1·83	1·75	1·67	1·57	1·52	1·46	1·39	1·32	1·22	1·00

ν_1, ν_2 are upper, lower d.f. respectively.

Table III   (cont.)

Upper 2·5% points

ν_2 \ ν_1	1	2	3	4	5	6	7	8	9	10	12	15	20	24	30	40	60	120	∞
1	647·8	799·5	864·2	899·6	921·8	937·1	948·2	956·7	963·3	968·6	976·7	984·9	993·1	997·2	1001	1006	1010	1014	1018
2	38·51	39·00	39·17	39·25	39·30	39·33	39·36	39·37	39·39	39·40	39·41	39·43	39·45	39·46	39·46	39·47	39·48	39·49	39·50
3	17·44	16·04	15·44	15·10	14·88	14·73	14·62	14·54	14·47	14·42	14·34	14·25	14·17	14·12	14·08	14·04	13·99	13·95	13·90
4	12·22	10·65	9·98	9·60	9·36	9·20	9·07	8·98	8·90	8·84	8·75	8·66	8·56	8·51	8·46	8·41	8·36	8·31	8·26
5	10·01	8·43	7·76	7·39	7·15	6·98	6·85	6·76	6·68	6·62	6·52	6·43	6·33	6·28	6·23	6·18	6·12	6·07	6·02
6	8·81	7·26	6·60	6·23	5·99	5·82	5·70	5·60	5·52	5·46	5·37	5·27	5·17	5·12	5·07	5·01	4·96	4·90	4·85
7	8·07	6·54	5·89	5·52	5·29	5·12	4·99	4·90	4·82	4·76	4·67	4·57	4·47	4·42	4·36	4·31	4·25	4·20	4·14
8	7·57	6·06	5·42	5·05	4·82	4·65	4·53	4·43	4·36	4·30	4·20	4·10	4·00	3·95	3·89	3·84	3·78	3·73	3·67
9	7·21	5·71	5·08	4·72	4·48	4·32	4·20	4·10	4·03	3·96	3·87	3·77	3·67	3·61	3·56	3·51	3·45	3·39	3·33
10	6·94	5·46	4·83	4·47	4·24	4·07	3·95	3·85	3·78	3·72	3·62	3·52	3·42	3·37	3·31	3·26	3·20	3·14	3·08
11	6·72	5·26	4·63	4·28	4·04	3·88	3·76	3·66	3·59	3·53	3·43	3·33	3·23	3·17	3·12	3·06	3·00	2·94	2·88
12	6·55	5·10	4·47	4·12	3·89	3·73	3·61	3·51	3·44	3·37	3·28	3·18	3·07	3·02	2·96	2·91	2·85	2·79	2·72
13	6·41	4·97	4·35	4·00	3·77	3·60	3·48	3·39	3·31	3·25	3·15	3·05	2·95	2·89	2·84	2·78	2·72	2·66	2·60
14	6·30	4·86	4·24	3·89	3·66	3·50	3·38	3·29	3·21	3·15	3·05	2·95	2·84	2·79	2·73	2·67	2·61	2·55	2·49
15	6·20	4·77	4·15	3·80	3·58	3·41	3·29	3·20	3·12	3·06	2·96	2·86	2·76	2·70	2·64	2·59	2·52	2·46	2·40
16	6·12	4·69	4·08	3·73	3·50	3·34	3·22	3·12	3·05	2·99	2·89	2·79	2·68	2·63	2·57	2·51	2·45	2·38	2·32
17	6·04	4·62	4·01	3·66	3·44	3·28	3·16	3·06	2·98	2·92	2·82	2·72	2·62	2·56	2·50	2·44	2·38	2·32	2·25
18	5·98	4·56	3·95	3·61	3·38	3·22	3·10	3·01	2·93	2·87	2·77	2·67	2·56	2·50	2·44	2·38	2·32	2·26	2·19
19	5·92	4·51	3·90	3·56	3·33	3·17	3·05	2·96	2·88	2·82	2·72	2·62	2·51	2·45	2·39	2·33	2·27	2·20	2·13
20	5·87	4·46	3·86	3·51	3·29	3·13	3·01	2·91	2·84	2·77	2·68	2·57	2·46	2·41	2·35	2·29	2·22	2·16	2·09
21	5·83	4·42	3·82	3·48	3·25	3·09	2·97	2·87	2·80	2·73	2·64	2·53	2·42	2·37	2·31	2·25	2·18	2·11	2·04
22	5·79	4·38	3·78	3·44	3·22	3·05	2·93	2·84	2·76	2·70	2·60	2·50	2·39	2·33	2·27	2·21	2·14	2·08	2·00
23	5·75	4·35	3·75	3·41	3·18	3·02	2·90	2·81	2·73	2·67	2·57	2·47	2·36	2·30	2·24	2·18	2·11	2·04	1·97
24	5·72	4·32	3·72	3·38	3·15	2·99	2·87	2·78	2·70	2·64	2·54	2·44	2·33	2·27	2·21	2·15	2·08	2·01	1·94
25	5·69	4·29	3·69	3·35	3·13	2·97	2·85	2·75	2·68	2·61	2·51	2·41	2·30	2·24	2·18	2·12	2·05	1·98	1·91
26	5·66	4·27	3·67	3·33	3·10	2·94	2·82	2·73	2·65	2·59	2·49	2·39	2·28	2·22	2·16	2·09	2·03	1·95	1·88
27	5·63	4·24	3·65	3·31	3·08	2·92	2·80	2·71	2·63	2·57	2·47	2·36	2·25	2·19	2·13	2·07	2·00	1·93	1·85
28	5·61	4·22	3·63	3·29	3·06	2·90	2·78	2·69	2·61	2·55	2·45	2·34	2·23	2·17	2·11	2·05	1·98	1·91	1·83
29	5·59	4·20	3·61	3·27	3·04	2·88	2·76	2·67	2·59	2·53	2·43	2·32	2·21	2·15	2·09	2·03	1·96	1·89	1·81
30	5·57	4·18	3·59	3·25	3·03	2·87	2·75	2·65	2·57	2·51	2·41	2·31	2·20	2·14	2·07	2·01	1·94	1·87	1·79
40	5·42	4·05	3·46	3·13	2·90	2·74	2·62	2·53	2·45	2·39	2·29	2·18	2·07	2·01	1·94	1·88	1·80	1·72	1·64
60	5·29	3·93	3·34	3·01	2·79	2·63	2·51	2·41	2·33	2·27	2·17	2·06	1·94	1·88	1·82	1·74	1·67	1·58	1·48
120	5·15	3·80	3·23	2·89	2·67	2·52	2·39	2·30	2·22	2·16	2·05	1·94	1·82	1·76	1·69	1·61	1·53	1·43	1·31
∞	5·02	3·69	3·12	2·79	2·57	2·41	2·29	2·19	2·11	2·05	1·94	1·83	1·71	1·64	1·57	1·48	1·39	1·27	1·00

ν_1, ν_2 are upper, lower d.f. respectively.

Table III (cont.)

Upper 1% points

v_2＼v_1	1	2	3	4	5	6	7	8	9	10	12	15	20	24	30	40	60	120	∞
1	4052	4999.5	5403	5625	5764	5859	5928	5981	6022	6056	6106	6157	6209	6235	6261	6287	6313	6339	6366
2	98.50	99.00	99.17	99.25	99.30	99.33	99.36	99.37	99.39	99.40	99.42	99.43	99.45	99.46	99.47	99.47	99.48	99.49	99.50
3	34.12	30.82	29.46	28.71	28.24	27.91	27.67	27.49	27.35	27.23	27.05	26.87	26.69	26.60	26.50	26.41	26.32	26.22	26.13
4	21.20	18.00	16.69	15.98	15.52	15.21	14.98	14.80	14.66	14.55	14.37	14.20	14.02	13.93	13.84	13.75	13.65	13.56	13.46
5	16.26	13.27	12.06	11.39	10.97	10.67	10.46	10.29	10.16	10.05	9.89	9.72	9.55	9.47	9.38	9.29	9.20	9.11	9.02
6	13.75	10.92	9.78	9.15	8.75	8.47	8.26	8.10	7.98	7.87	7.72	7.56	7.40	7.31	7.23	7.14	7.06	6.97	6.88
7	12.25	9.55	8.45	7.85	7.46	7.19	6.99	6.84	6.72	6.62	6.47	6.31	6.16	6.07	5.99	5.91	5.82	5.74	5.65
8	11.26	8.65	7.59	7.01	6.63	6.37	6.18	6.03	5.91	5.81	5.67	5.52	5.36	5.28	5.20	5.12	5.03	4.95	4.86
9	10.56	8.02	6.99	6.42	6.06	5.80	5.61	5.47	5.35	5.26	5.11	4.96	4.81	4.73	4.65	4.57	4.48	4.40	4.31
10	10.04	7.56	6.55	5.99	5.64	5.39	5.20	5.06	4.94	4.85	4.71	4.56	4.41	4.33	4.25	4.17	4.08	4.00	3.91
11	9.65	7.21	6.22	5.67	5.32	5.07	4.89	4.74	4.63	4.54	4.40	4.25	4.10	4.02	3.94	3.86	3.78	3.69	3.60
12	9.33	6.93	5.95	5.41	5.06	4.82	4.64	4.50	4.39	4.30	4.16	4.01	3.86	3.78	3.70	3.62	3.54	3.45	3.36
13	9.07	6.70	5.74	5.21	4.86	4.62	4.44	4.30	4.19	4.10	3.96	3.82	3.66	3.59	3.51	3.43	3.34	3.25	3.17
14	8.86	6.51	5.56	5.04	4.69	4.46	4.28	4.14	4.03	3.94	3.80	3.66	3.51	3.43	3.35	3.27	3.18	3.09	3.00
15	8.68	6.36	5.42	4.89	4.56	4.32	4.14	4.00	3.89	3.80	3.67	3.52	3.37	3.29	3.21	3.13	3.05	2.96	2.87
16	8.53	6.23	5.29	4.77	4.44	4.20	4.03	3.89	3.78	3.69	3.55	3.41	3.26	3.18	3.10	3.02	2.93	2.84	2.75
17	8.40	6.11	5.18	4.67	4.34	4.10	3.93	3.79	3.68	3.59	3.46	3.31	3.16	3.08	3.00	2.92	2.83	2.75	2.65
18	8.29	6.01	5.09	4.58	4.25	4.01	3.84	3.71	3.60	3.51	3.37	3.23	3.08	3.00	2.92	2.84	2.75	2.66	2.57
19	8.18	5.93	5.01	4.50	4.17	3.94	3.77	3.63	3.52	3.43	3.30	3.15	3.00	2.92	2.84	2.76	2.67	2.58	2.49
20	8.10	5.85	4.94	4.43	4.10	3.87	3.70	3.56	3.46	3.37	3.23	3.09	2.94	2.86	2.78	2.69	2.61	2.52	2.42
21	8.02	5.78	4.87	4.37	4.04	3.81	3.64	3.51	3.40	3.31	3.17	3.03	2.88	2.80	2.72	2.64	2.55	2.46	2.36
22	7.95	5.72	4.82	4.31	3.99	3.76	3.59	3.45	3.35	3.26	3.12	2.98	2.83	2.75	2.67	2.58	2.50	2.40	2.31
23	7.88	5.66	4.76	4.26	3.94	3.71	3.54	3.41	3.30	3.21	3.07	2.93	2.78	2.70	2.62	2.54	2.45	2.35	2.26
24	7.82	5.61	4.72	4.22	3.90	3.67	3.50	3.36	3.26	3.17	3.03	2.89	2.74	2.66	2.58	2.49	2.40	2.31	2.21
25	7.77	5.57	4.68	4.18	3.85	3.63	3.46	3.32	3.22	3.13	2.99	2.85	2.70	2.62	2.54	2.45	2.36	2.27	2.17
26	7.72	5.53	4.64	4.14	3.82	3.59	3.42	3.29	3.18	3.09	2.96	2.81	2.66	2.58	2.50	2.42	2.33	2.23	2.13
27	7.68	5.49	4.60	4.11	3.78	3.56	3.39	3.26	3.15	3.06	2.93	2.78	2.63	2.55	2.47	2.38	2.29	2.20	2.10
28	7.64	5.45	4.57	4.07	3.75	3.53	3.36	3.23	3.12	3.03	2.90	2.75	2.60	2.52	2.44	2.35	2.26	2.17	2.06
29	7.60	5.42	4.54	4.04	3.73	3.50	3.33	3.20	3.09	3.00	2.87	2.73	2.57	2.49	2.41	2.33	2.23	2.14	2.03
30	7.56	5.39	4.51	4.02	3.70	3.47	3.30	3.17	3.07	2.98	2.84	2.70	2.55	2.47	2.39	2.30	2.21	2.11	2.01
40	7.31	5.18	4.31	3.83	3.51	3.29	3.12	2.99	2.89	2.80	2.66	2.52	2.37	2.29	2.20	2.11	2.02	1.92	1.80
60	7.08	4.98	4.13	3.65	3.34	3.12	2.95	2.82	2.72	2.63	2.50	2.35	2.20	2.12	2.03	1.94	1.84	1.73	1.60
120	6.85	4.79	3.95	3.48	3.17	2.96	2.79	2.66	2.56	2.47	2.34	2.19	2.03	1.95	1.86	1.76	1.66	1.53	1.38
∞	6.63	4.61	3.78	3.32	3.02	2.80	2.64	2.51	2.41	2.32	2.18	2.04	1.88	1.79	1.70	1.59	1.47	1.32	1.00

v_1, v_2 are upper, lower d.f. respectively.

Table III (cont.)

Upper 0·1% points

ν_2 \ ν_1	1	2	3	4	5	6	7	8	9	10	12	15	20	24	30	40	60	120	∞
1	4053*	5000*	5404*	5625*	5764*	5859*	5929*	5981*	6023*	6056*	6107*	6158*	6209*	6235*	6261*	6287*	6313*	6340*	6366*
2	998·5	999·0	999·2	999·2	999·3	999·3	999·4	999·4	999·4	999·4	999·4	999·4	999·4	999·5	999·5	999·5	999·5	999·5	999·5
3	167·0	148·5	141·1	137·1	134·6	132·8	131·6	130·6	129·9	129·2	128·3	127·4	126·4	125·9	125·4	125·0	124·5	124·0	123·5
4	74·14	61·25	56·18	53·44	51·71	50·53	49·66	49·00	48·47	48·05	47·41	46·76	46·10	45·77	45·43	45·09	44·75	44·40	44·05
5	47·18	37·12	33·20	31·09	29·75	28·84	28·16	27·64	27·24	26·92	26·42	25·91	25·39	25·14	24·87	24·60	24·33	24·06	23·79
6	35·51	27·00	23·70	21·92	20·81	20·03	19·46	19·03	18·69	18·41	17·99	17·56	17·12	16·89	16·67	16·44	16·21	15·99	15·75
7	29·25	21·69	18·77	17·19	16·21	15·52	15·02	14·63	14·33	14·08	13·71	13·32	12·93	12·73	12·53	12·33	12·12	11·91	11·70
8	25·42	18·49	15·83	14·39	13·49	12·86	12·40	12·04	11·77	11·54	11·19	10·84	10·48	10·30	10·11	9·92	9·73	9·53	9·33
9	22·86	16·39	13·90	12·56	11·71	11·13	10·70	10·37	10·11	9·89	9·57	9·24	8·90	8·72	8·55	8·37	8·19	8·00	7·81
10	21·04	14·91	12·55	11·28	10·48	9·92	9·52	9·20	8·96	8·75	8·45	8·13	7·80	7·64	7·47	7·30	7·12	6·94	6·76
11	19·69	13·81	11·56	10·35	9·58	9·05	8·66	8·35	8·12	7·92	7·63	7·32	7·01	6·85	6·68	6·52	6·35	6·17	6·00
12	18·64	12·97	10·80	9·63	8·89	8·38	8·00	7·71	7·48	7·29	7·00	6·71	6·40	6·25	6·09	5·93	5·76	5·59	5·42
13	17·81	12·31	10·21	9·07	8·35	7·86	7·49	7·21	6·98	6·80	6·52	6·23	5·93	5·78	5·63	5·47	5·30	5·14	4·97
14	17·14	11·78	9·73	8·62	7·92	7·43	7·08	6·80	6·58	6·40	6·13	5·85	5·56	5·41	5·25	5·10	4·94	4·77	4·60
15	16·59	11·34	9·34	8·25	7·57	7·09	6·74	6·47	6·26	6·08	5·81	5·54	5·25	5·10	4·95	4·80	4·64	4·47	4·31
16	16·12	10·97	9·00	7·94	7·27	6·81	6·46	6·19	5·98	5·81	5·55	5·27	4·99	4·85	4·70	4·54	4·39	4·23	4·06
17	15·72	10·66	8·73	7·68	7·02	6·56	6·22	5·96	5·75	5·58	5·32	5·05	4·78	4·63	4·48	4·33	4·18	4·02	3·85
18	15·38	10·39	8·49	7·46	6·81	6·35	6·02	5·76	5·56	5·39	5·13	4·87	4·59	4·45	4·30	4·15	4·00	3·84	3·67
19	15·08	10·16	8·28	7·26	6·62	6·18	5·85	5·59	5·39	5·22	4·97	4·70	4·43	4·29	4·14	3·99	3·84	3·68	3·51
20	14·82	9·95	8·10	7·10	6·46	6·02	5·69	5·44	5·24	5·08	4·82	4·56	4·29	4·16	4·00	3·86	3·70	3·54	3·38
21	14·59	9·77	7·94	6·95	6·32	5·88	5·56	5·31	5·11	4·95	4·70	4·44	4·17	4·03	3·88	3·74	3·58	3·42	3·26
22	14·38	9·61	7·80	6·81	6·19	5·76	5·44	5·19	4·99	4·83	4·58	4·33	4·06	3·92	3·78	3·63	3·48	3·32	3·15
23	14·19	9·47	7·67	6·69	6·08	5·65	5·33	5·09	4·89	4·73	4·48	4·23	3·96	3·82	3·68	3·53	3·38	3·22	3·05
24	14·03	9·34	7·55	6·59	5·98	5·55	5·23	4·99	4·80	4·64	4·39	4·14	3·87	3·74	3·59	3·45	3·29	3·14	2·97
25	13·88	9·22	7·45	6·49	5·88	5·46	5·15	4·91	4·71	4·56	4·31	4·06	3·79	3·66	3·52	3·37	3·22	3·06	2·89
26	13·74	9·12	7·36	6·41	5·80	5·38	5·07	4·83	4·64	4·48	4·24	3·99	3·72	3·59	3·44	3·30	3·15	2·99	2·82
27	13·61	9·02	7·27	6·33	5·73	5·31	5·00	4·76	4·57	4·41	4·17	3·92	3·66	3·52	3·38	3·23	3·08	2·92	2·75
28	13·50	8·93	7·19	6·25	5·66	5·24	4·93	4·69	4·50	4·35	4·11	3·86	3·60	3·46	3·32	3·18	3·02	2·86	2·69
29	13·39	8·85	7·12	6·19	5·59	5·18	4·87	4·64	4·45	4·29	4·05	3·80	3·54	3·41	3·27	3·12	2·97	2·81	2·64
30	13·29	8·77	7·05	6·12	5·53	5·12	4·82	4·58	4·39	4·24	4·00	3·75	3·49	3·36	3·22	3·07	2·92	2·76	2·59
40	12·61	8·25	6·60	5·70	5·13	4·73	4·44	4·21	4·02	3·87	3·64	3·40	3·15	3·01	2·87	2·73	2·57	2·41	2·23
60	11·97	7·76	6·17	5·31	4·76	4·37	4·09	3·87	3·69	3·54	3·31	3·08	2·83	2·69	2·55	2·41	2·25	2·08	1·89
120	11·38	7·32	5·79	4·95	4·42	4·04	3·77	3·55	3·38	3·24	3·02	2·78	2·53	2·40	2·26	2·11	1·95	1·76	1·54
∞	10·83	6·91	5·42	4·62	4·10	3·74	3·47	3·27	3·10	2·96	2·74	2·51	2·27	2·13	1·99	1·84	1·66	1·45	1·00

* Multiply these entries by 100.

ν_1, ν_2 are upper, lower d.f. respectively.

Table IV   Values of the correlation coefficient, ρ, which differ significantly from 0 at the 5%, 1%, 0·1% levels.

d.f.	0·05	0·01	0·001	d.f.	0·05	0·01	0·001
1	0·9²692	0·9³877	0·9⁵877	16	0·468	0·590	0·708
2	·9500	·9²000	·9³000	17	·456	·575	·693
3	·878	·9587	·9³114	18	·444	·561	·679
4	·811	·9172	·9741	19	·433	·549	·665
5	·754	·875	·9509	20	·423	·537	·652
6	0·707	0·834	0·9249	25	0·381	0·487	0·597
7	·666	·798	·898	30	·349	·449	·554
8	·632	·765	·872	35	·325	·418	·519
9	·602	·735	·847	40	·304	·393	·490
10	·576	·708	·823	45	·288	·372	·465
11	0·553	0·684	0·801	50	0·273	0·354	0·443
12	·532	·661	·780	60	·250	·325	·408
13	·514	·641	·760	70	·232	·302	·380
14	·497	·623	·742	80	·217	·283	·357
15	·482	·606	·725	90	·205	·267	·338
				100	·195	·254	·321

The foregoing Tables are reprinted, by permission of the Trustees, from *Biometrika Tables for Statisticians*, 3rd Edition (1966), ed. E. S. Pearson and H. O. Hartley.

Table V   Values of Spearman's rank correlation coefficient r_s which differ significantly from 0 at the 5% and 1% levels using a two-tail test.

Sample size N	5% level	1% level
5	1·000	–
6	0·886	1·000
7	0·750	0·893
8	0·714	0·857
9	0·683	0·833
10	0·648	0·794

For sample sizes greater than $N = 10$ use Table IV.

Index

I: STATISTICAL TERMS

Additivity, 103, 112
Alternative Hypothesis, 53–5, 106, 151
Analysis of variance, 94–6, 103–7, 116–18, 132–3, 137–9, 145–6
assumptions, 111–12, 146
Angular transformation, 147
Average, xii, 15

Bar-chart, 8, 9, 29
Bartlett, 112, 148, 158
Binomial distribution, 24–30, 35, 40, 43, 53, 67, 75, 79, 147, 150–1
negative, 35–6
tests in, 53, 67, 75
transformation of, 147
Block, 114–19
incomplete, 122
randomized, 114–19, 136–9, 143–4, 152

Central Limit Theorem, 41, 51, 63–4, 149
Chi-squared, 64–73, 81–2, 149
Class and class-interval, 9–10
Coefficient:
of variation, 127
(see also Correlation, Regression)
Completely Randomized design, 103–11, 145–6

Contingency table, 68–9, 71–3
Continuous variate (see Variate)
Control treatment, 109
Correction, Yates', 69
Correlation and correlation coefficient, 83–91
Correlation, rank, 88–9
Covariance, 84
Cumulative:
frequency, 8, 11, 47
probability, 39
Curve:
frequency, 13, 14, 38
response, 134–5

Degrees of freedom, 58, 64, 73, 124–5
(see also Analysis of variance)
Density function, probability, 13
Dependent variate in regression, 92, 99–100
Design (see Completely Randomized, Randomized Block, Latin Square, Graeco-Latin Square)
Deviation:
mean, 21
standard, 21, 41–2
Diagram (see Bar-chart, Histogram, Scatter diagram)

Difference, significant, 108, 118, 121, 124, 126
Discrete variate (*see* Variate)
Dispersion, 14, 18–21
index of, 66–7
Distribution, xii
function, 13
sampling, 41–2
(*see also* Binomial, Normal, Lognormal, Poisson)
Duncan's test, 110

Ecology, distributions in, 33, 36
Effect, main, 130–3, 137–9
Error:
experimental, xi, 103
standard, of treatment means and differences, 109–10
Estimation, 29–30, 35

F-test, 73–4, 96, 106–7, 116–17, 121, 132–3, 138
Factor, 128
Factorial experiment, 109, 110, 128–41
Fit, goodness of, 70
for regression line, 94–6
Freedom, *see* Degrees of freedom
Frequency, 7
cumulative, 8, 11, 47
curve, 13, 14, 37–9
distribution function, 13
polygon, 12, 13
table, 7–9, 11
Function:
frequency distribution, 13
probability density, 13

Gaussian distribution (*see* Normal)
Goodness of fit (*see* Fit)
Graeco-Latin Square, 122

Histogram, 10
Hypothesis, 44
Alternative, 53–5, 106, 151
Null, 46, 97, 106

Incomplete blocks, 122
Independence, 33
Independent variate, *see* Variate
Index of dispersion, 66–7
Interaction, 115, 128–39

Latin Square, 119–22, 144
Least significant difference (*see* Significant difference)
Least squares, 93
Level of significance, 48
Limit (*see* Central Limit Theorem)
Limits, 77–82, 97
approximate, 79
large-sample, 81
normal mean, 77–8
normal variance, 81–2
Poisson mean, 80
proportion, 80
regression coefficient, 97
Linear regression, 92–7
Location, 14–18, 149–51
(*see also* Mean, Median, Mode)
Logarithmic transformation, 147
Log-normal distribution, 40, 149

Main effect, 130–3, 137–9
Matched pairs, 152–4
Mean, 14–16, 38, 41, 42, 149
(*see also* Limits, Significance test)
deviation, 21
square, 96, 107, 116, 121
standard error of, 109–10
Median, 17–18, 41, 150–5
Missing observations (*or* plots), 142–5
Mixed factorial, 135–41
Mode, 17

Natural variation, xi–xii
Negative binomial, 35–6
Nonlinear regression, 97–9
Nonparametric test of significance, 149–56
Normal distribution, 37–57, 77–8, 81–2, 97, 151, 153–4
limits to mean, 77–8

Normal distribution—*cont.*
  limits to variance, 81–2
  standard (*or* unit), 39, 46–8, 58
  (*see also* Analysis of variance,
    Linear regression)
Normality, approximate, 51–3, 79–
    81, 149
Null Hypothesis, 46, 97, 106
Numbers, random, 76, 111, 115

Observation, missing, 142–5
One-tail test of significance, 53–5,
    64–5, 70, 73, 96, 151
Orthogonality, 111, 118–19

Paired samples, 151–4
Pairs, matched, 152–4
Parameter, 28, 32, 38
Permutation, random, 111, 115
Plots, 39, 101
  missing, 142–5
Poisson distribution, 32–5, 41, 56,
    66–7, 80, 147
  limits to mean, 80
  tests in, 52, 66–7, 75
  transformation of, 147
Polygon, frequency, 12, 13
Population, xi, 1
Precision, 124–7
Probability, 23–8
  cumulative, 39
  density function, 13
  prior, 45
Process, stochastic, 33
Proportions:
  limits, 80
  significance tests of, 53, 68–9

Qualitative variate, 3
Quantitative variate, 3

Randomization, 103
Randomized Block, 114–19, 136–9,
    143–4, 152
Randomized design, Completely,
    103–11, 145–6
Randomness, 33, 36

Random:
  numbers, 76, 111, 115
  permutation, 111, 115
  sample, 3–6
Range, 21
Range-test (multiple), 110
Rank correlation, 88–9
Ranking, 88–91, 150–6
Ranks, tied, 153
Regression:
  coefficient (slope), 93–4
    limits to, 97
  linear, 92–7
    goodness of fit, 94–6
    significance tests in, 95–6
  nonlinear, 97–9
Replacement, sampling with or with-
    out, 4–5
Replication, 102, 124
Response curve, 134–5
Robust, 149–50, 155

Sample, 2
  paired, 151–4
  (*see also* Random, Replacement)
Sampling distribution, 41–2
Scatter, 14
  diagram, 85
Sign test, 150–2
Signed-ranks test (Wilcoxon), 152–4
Significance level, 48
Significance test, 40, 44, 157–8
  chi-squared, 64–73, 149
  F, 73–4, 96, 106–7, 116–17, 121,
    132–3, 138
  for means, 50–2, 54–5, 59–63,
    107–8, 133, 149
  for proportions, 53, 68–9
  for variances, 64–6, 73–4
  in binomial distribution, 53, 66–8,
    75
  in normal distribution, 46–51,
    54–5, 64–6, 73–4
  in regression, 95–6
  nonparametric, 149–56
  normal approximation, 51–3, 151,
    153–4

Significance test—*cont.*
(*see also* Analysis of variance, Duncan, Index of dispersion, One-tail, Range, Sign, Signed-ranks, *t*, Two-tail, U-test)
Significant difference (least), 108, 118, 121, 124, 126
Skewness, 13, 14
Slope of regression line, 93–4
limits to, 97
Spearman, 88–9
Square (*see* Graeco-Latin, Latin, Mean)
Squares:
least, 93
sum of, 94, 104, 116, 120
Standard:
deviation, 21, 41–2
error, 109–10
normal distribution, 39, 46–8, 58
Stochastic process, 33
Student's *t* (*see* *t*)
Survey, 2
Systematic variation, 114–16, 119, 125

t, Student's, 58–63, 77–9, 97, 107–8, 124, 133, 139, 146, 149–50
Table (*see* Contingency, Frequency)
Theorem (*see* Central Limit)
Tied ranks, 153

Transformation, 40, 98, 147
Two-tail test of significance, 53–5, 64–6, 74, 151

U-test, 154–5
Unit normal distribution, 39, 46–8, 58

Variance, 14, 18–21
limits to, 81–2
significance tests for, 64–6, 73–4
(*see also* Analysis of variance)
Variate, 3
continuous, 3, 7, 9, 37–51, 54–5, 58–66, 73–4, 77–9, 81–2
dependent, 92, 99–100
discrete, 3, 7, 8, 23–36, 52–3, 66–73, 79–80
independent, 92, 99–100
qualitative, 3
quantitative, 3
Variation:
coefficient of, 127
(*see* Natural, Systematic)

Wilcoxon (*see* Signed-ranks test)

χ^2, 64–73, 81–2, 149

Yates' correction, 69

II: TOPICS USED AS EXAMPLES

Accidents, numbers of, 36
Animals:
diets, 131–4
weights of, 11, 21, 56, 75, 131–4, 143–4
Aptitude tests, 120–2

Bacterial suspension, counts in, 117–18

Blood-clotting time, 90

Cells, counts, 32–4, 82
Chemical analysis, 86–7, 90, 112–13, 135–9
Children, weights, 57
Chromosomes, 30, 49–50, 56
Coin tossing, 23
Colour of leaf, 71–2

Colouring agents, food, 153
Counts (*see* Accidents, Bacterial suspension, Cells, Eggs, Insects, Particles, Plants, Telephone calls)
Crop yield (weight), 21–2, 56, 61, 86, 100, 105–6, 108, 122, 127, 140

Density, optical, 92–7
Die, throws of, 7, 21
Diets of animals, 131–4
Drugs, 56, 75–6, 82

Eggs, hens', weights, 22
Eggs of insects, counts of, 35

Fertility trend, soil, 114–15, 119
Food:
  colouring, 153
  packaging, 31
  storage, 155–6
Fungicide, 123, 140, 152
Fungus, 51, 54–5, 75

Genotypes, 24, 28, 30, 70, 76, 82
Germination of seeds, 30, 68–9

Heights:
  human, 90
  plant, 7, 9–11, 12, 16, 75–6, 82, 113, 127, 140–1

Insects:
  counts, 33, 35
  eggs, counts, 35

Jam, colour and texture, 88–9

Leaf:
  chemistry, 86–7, 135–9
  colour, 71–2

Male rats, number per litter, 11, 15–16, 19–20, 29, 67

Optical density, 92–7

Packaging:
  fertilizer, 43
  food, 31
Particles, radioactive, 8, 17, 21, 33, 35–6, 43, 52, 67
Plant:
  breeding, 30, 53
  counts, 33
  heights (*see* Heights)
  vigour, 71–2
  weights, 76, 155
Playing cards, 11

Quadrats, sampling, 21–2, 36, 56

Radioactive particles (*see* Particles)
Rainfall, 90
Rats (*see* Male rats)
Reaction time, 155

Salaries, 100
Sampling quadrats, 21–2, 36, 56
Seeds, germination, 30, 68–9
Soil:
  conditioner, 56
  type, 76
Sprays, 123, 140, 148
Stimulus, in psychological test, 155
Storage, food, 155–6
Suspension, bacterial, 117–18

Tasting, 123, 155–6
Telephone calls, counts, 33
Test, aptitude, 120–2
  tasting, 123
Throws of die, 7, 21
Tossing, coin-, 23
Trend, fertility, 114–15, 119

Vigour of plants, 71–2

Weights (*see* Animals, Plant, Eggs, Children, Crop)

Yields (*see* Crop)